London Mathematical Society Student Texts 112

The Shrikhande Graph

A Window on Discrete Mathematics

PETER J. CAMERON
University of St Andrews

APARNA LAKSHMANAN S.
Cochin University of Science and Technology

AMBAT VIJAYAKUMAR
Cochin University of Science and Technology

Shaftesbury Road, Cambridge CB2 8EA, United Kingdom

One Liberty Plaza, 20th Floor, New York, NY 10006, USA

477 Williamstown Road, Port Melbourne, VIC 3207, Australia

314–321, 3rd Floor, Plot 3, Splendor Forum, Jasola District Centre,
New Delhi – 110025, India

Cambridge University Press is part of Cambridge University Press & Assessment,
a department of the University of Cambridge.

We share the University's mission to contribute to society through the pursuit of
education, learning and research at the highest international levels of excellence.

www.cambridge.org
Information on this title: www.cambridge.org/9781009709088
DOI: 10.1017/9781009709118

© Peter J. Cameron, Aparna Lakshmanan S. and Ambat Vijayakumar 2026

This publication is in copyright. Subject to statutory exception and to the provisions of
relevant collective licensing agreements, no reproduction of any part may take place
without the written permission of Cambridge University Press & Assessment.

When citing this work, please include a reference to the
DOI 10.1017/9781009709118

First published 2026

A catalogue record for this publication is available from the British Library

*A Cataloging-in-Publication data record for this book is available from the Library of
Congress*

ISBN 978-1-009-70910-1 Hardback
ISBN 978-1-009-70908-8 Paperback

Cambridge University Press & Assessment has no responsibility for the persistence
or accuracy of URLs for external or third-party internet websites referred to in this
publication and does not guarantee that any content on such websites is, or will
remain, accurate or appropriate.

For EU product safety concerns, contact us at Calle de José Abascal, 56, 1°, 28003 Madrid,
Spain, or email eugpsr@cambridge.org.

Contents

Preface

The *Petersen graph*, a strongly regular graph on 10 vertices, is so famous, and its occurrences within graph theory as an example or a counterexample are so many and varied, that it has a whole book devoted to it [59].

Our subject here is the *Shrikhande graph*, discovered by the Indian mathematician Sharadchandra Shankar Shrikhande in 1959. Our motivation is a little different to that of Derek Holton and John Sheehan. The Shrikhande graph occupies a pivotal position within discrete mathematics, at the meeting point of algebraic and topological graph theory, Latin squares, root systems, graph eigenvalues, Seidel switching and other topics. We invite the reader to a gentle introduction to these topics.

In this book, the unadorned term 'the Graph' will sometimes be used to refer to the Shrikhande graph.

If you look at the picture of the Shrikhande graph on the torus (Figure 7.1), you will see that from a vertex there are six different directions you can travel. In this book, we are going to give six different constructions of the Shrikhande graph:

(a) a direct construction (Section 3.5), which can be used to calculate the number of automorphisms (Section 4.4);
(b) as a Cayley graph (Sections 2.6 and 4.3);
(c) as the complement of the Latin square graph from the Cayley table of the cyclic group of order 4 (Section 6.7);
(d) as a regular map on the torus (Section 7.3);
(e) embedded in the exceptional root system E_7 (Section 9.3);
(f) and by Seidel switching from the 4×4 grid graph, the line graph of $K_{4,4}$ (Section 10.2).

Each construction is embedded in a detailed account of the relevant part of discrete mathematics.

The book begins with a short biography of Shrikhande. He is best known as one of the 'Euler spoilers', the trio of mathematicians who disproved the conjecture of Euler on the existence of Graeco-Latin squares. (For the novice, we remark that the name 'Euler' is pronounced to rhyme with 'spoiler'.) In keeping with this, we observe that there are two Latin squares of order 4, the Cayley tables of the two groups of order 4; the complements of their Latin square graphs are $L_2(4)$ (the line graph of the complete bipartite graph $K_{4,4}$) and the Shrikhande graph.

This is followed by two chapters giving an introduction to the notation and terminology of graph theory and to the class of strongly regular graphs (including the Shrikhande graph and several other beautiful graphs). The concepts introduced are illustrated in Chapters 1 and 2 by showing how they work out in the case of the Shrikhande graph.

The final part of the book was one of our main reasons for writing it: a discussion of the various contexts in which the Shrikhande graph arises, treating Latin squares, triangulations of the torus, root systems, Seidel switching and others. Many of these contexts throw further light on the properties of the Shrikhande graph and its special place in graph theory.

Thus, we use the Shrikhande graph as a window for viewing a number of topics in discrete mathematics, including strongly regular and distance-regular graphs, Latin squares, generalized line graphs, error-correcting codes and several others.

Shrikhande's theorem states that a graph with the same spectrum as the line graph of the complete bipartite graph $K_{n,n}$ is isomorphic to $L(K_{n,n})$ unless $n = 4$, in which case there is just one further graph, the Shrikhande graph. Two extensions of this theorem we discuss are classifications of connected graphs with smallest eigenvalue -2 (where the Shrikhande graph is one of the finitely many exceptions to the statement that these are generalized line graphs) and distance-regular graphs with the same parameters as Hamming graphs (where the Doob graphs, Cartesian products of 4-cliques and copies of the Shrikhande graph, are the exceptions when the alphabet size is 4). Also, a set of mutually orthogonal Latin squares containing two fewer than a complete set can be completed if the order is not 4; the exception for order 4 is explained by the Shrikhande graph.

Acknowledgements

Shrikhande's discovery of his graph was not a one-off event, but was part of a much wider project at the time, to characterize certain graphs by the spectra of their adjacency matrices. The American mathematician Alan Hoffman was a leading figure in this project, but it also involved mathematicians from many countries, including Belgium, China, India, the Netherlands, the United Kingdom and the United States. It will also be made clear from this book that the Shrikhande graph is connected to a wide range of topics in discrete mathematics, a worldwide language and enterprise.

The subject has a rich history. Some of this, such as Euler's conjecture on Graeco-Latin squares, we treat in detail; but we also give references to other interesting byways, including Mesner's discovery of the Higman–Sims graph, Paley's non-discovery of the Paley graphs and the story of Hoffman's bound, for which we recommend the papers [65], [61] and [55], respectively, by Klin and Woldar, Jones and Haemers, respectively. We are grateful to all these authors for their papers and discussion.

We have attempted to give references for further reading on the topics we cover, including the surprising occurrence of the Coxeter–Dynkin diagrams of type ADE in the theory of graph spectra.

We are grateful to many colleagues and friends who have provided us with the information which we are passing on in these pages. In particular, we thank Sharad Sane and Mohan Shrikhande for some biographical information about S. S. Shrikhande, Andries Brouwer for the construction of a locally Shrikhande graph, Wilfried Imrich for results about the primality of the Shrikhande graph and R. A. Bailey for the anecdote in Chapter 1.

The pictures in Chapter 6 and Section 10.6 are from photographs by the first author, while those in Sections 6.2 and 10.7 are © Neill Cameron. In other cases, we have attempted to use pictures in the public domain.

PART I

Biography

1

The Life of S. S. Shrikhande

Ambat Vijayakumar, an author of this book, writes:

The International Conference on Recent Trends in Graph Theory and Combinatorics (ICRTGC) was held in Cochin, India. It was organized by the Cochin University of Science and Technology, India, during 7–10 June 2010, as a satellite conference of the International Congress of Mathematicians (ICM) 2010 held in Hyderabad, India. The conference logo (Fig. 1.1) was the renowned 'Shrikhande Graph'. I had only a vague memory of having met S. S. Shrikhande in a conference held at the University of Mumbai and I had never heard about his contributions to combinatorial designs, association schemes, and the Shrikhande Graph itself before 2010. Although I do not wish to find excuses for my ignorance, it is surprising that the Shrikhande Graph came to my mind. That prompted me to know more about him and his graph, and from then I was

Figure 1.1 Logo of ICRTGC 2010

Figure 1.2 Sharadchandra Shankar Shrikhande

madly popularizing the works of Shrikhande to the mathematical community
through webinars, articles for students in my mother tongue (Malayalam), and
so forth.

This introductory chapter provides a biographical sketch of Sharadchandra
Shankar Shrikhande. Shrikhande (Fig. 1.2) was born at Sagar, Madhya Pradesh,
on 19 October 1917, as the fifth of ten children to his parents Shankar and
Parvathy. Like many other Indian mathematicians, he was not born with a sil-
ver spoon; his father being an ordinary worker at a flour mill. He received his
basic education at Government High School, Sagar, and passed the school-
leaving certificate examination in 1934 with the fifth rank. Consequently, he
was awarded a government merit scholarship to continue his studies at Robertson College, Jabalpur, for the intermediate science course, which he completed
with flying colours in 1936. This success enabled him to get another scholarship to study the BSc Honours course at the Government College of Science,
formerly known as the Institute of Science, Nagpur (since 1906). He completed
this course majoring in pure and applied mathematics, with physics as an allied
subject, securing the first rank, for which he was awarded the Prakya Ganpat
Rao Gold Medal. He was very badly in need of employment, which forced him
to move to Calcutta (now Kolkata) in West Bengal, India.

The story of Shrikhande now has to be linked with that of Professor
P. C. Mahalanobis (1893–1972), the father of modern Indian statistics and
the prestigious Indian Statistical Institute (ISI), which he established in 1930.
Though Mahalanobis was a Professor of Physics at the Presidency College,
Calcutta, he made remarkable contributions to statistical techniques, in jute
sample surveys. Shrikhande joined the ISI in 1940 with financial help, mainly

through the King Edward Memorial Fellowship, named after Edward VIII (1894–1972), who was King of the United Kingdom and the Dominions of the British Empire and Emperor of India in 1936. The ISI already had the presence of S. N. Roy, R. C. Bose, K. R. Nair (who later became an FRS), and many others, who all later became doyens in statistics due to their fundamental contributions. He returned to his alma mater at Jabalpur, and later moved to Government College of Science at Nagpur, to work as a lecturer from 1942 to 1958. But he was making regular visits to the ISI to have discussions with R. C. Bose, who introduced him to the theory of statistical designs, Latin squares, and similar topics.

In 1947, on the suggestion of Dr. P. V. Sukhathme (the Statistical Adviser of Indian Council of Agricultural Research, New Delhi), Shrikhande joined the Department of Mathematics and Statistics at the University of North Carolina, USA. In fact, the Government of India had decided to send some Indian scholars to the USA for training under the Public Law 480 – Agricultural Trade Development Assistance Act scheme. That department had just been established, in 1946, under the chairmanship of Professor Harold Hotelling (an American statistician well known for Hotelling's law as well as Hotelling's T-squared distribution) and already had the services of K. A. Bush, R. A. Bradley, R. C. Bose, S. N. Roy, and G. Kallianpur (who later became Director of the ISI).

Shrikhande's formal studies began with a course on Linear Estimation by R. C. Bose. Later he became Bose's first PhD student, and completed his thesis in 1950. Shrikhande describes the return of Bose to the USA from India in 1949 as a 'turning point in my academic career'. K. A. Bush, who later served Washington State University, USA, remarks in *A Survey of Combinatorial Theory* (1973), pp. 79–80: 'Professor Bose set highly imaginative but quite difficult topics for the student's investigation. In the second place, the student had to supply substantial originality. Professor Bose seemed totally unaware of the modern trend where the thesis advisor shows the student how to solve a part of the problem and then suggests that he carry it further with some inconsequential generalization. Instead, Professor Bose expected students to discover appropriate theorems on their own and develop sound techniques for their proof. At this stage, he would read every word, frequently rewrite entire sections to improve the exposition, and sometimes find important refinements or extensions of the results. While this was helpful, the really important element is that he forced the students to do genuine research from the outset. Professor Bose made the students feel that his overwhelming desire was to see each of the students succeed.'

During 1951–53, Shrikhande was an Assistant Professor of Statistics at the University of Kansas, Lawrence, USA, and was Associate Professor at Chapel

Hill during 1958–60. It was during the latter period that the famous refutation of Euler's conjecture on Graeco-Latin squares (which we discuss later in the book) was given by Bose, Shrikhande, and E. T. Parker. Instead of staying on in the USA, Shrikhande returned to India to take up a professorship at Banaras Hindu University, where he worked until 1963. He then joined the University of Bombay as Professor and Head of the Department of Mathematics, where he worked until his formal retirement in 1978. During this period, he was also Director of the Centre for Advanced Study in Mathematics at Bombay. During 1983–86, he was Director of the Mehta Research Institute at Allahabad, which is a research centre for Mathematics and Theoretical Physics (renamed later as Harish Chandra Research Institute).

He had been a Visiting Professor at various US universities, such as the University of Wisconsin, the Ohio State University, State University of New York, Stanford University, and Colorado State University. He has been associated with the Indian Statistical Institute in various capacities. Professor Shrikhande has been a member of a number of reputed societies: Indian National Science Academy, Indian Academy of Sciences, Institute of Mathematical Statistics, and International Statistical Institute, to mention a few.

Shrikhande spent a few years in Nagpur after leaving his position as the Director of the Mehta Research Institute. His last ten years passed very peacefully and were spent in the extremely quiet and serene surroundings of Chinmaya Vijay Ashram, Vijayawada, in Andhra Pradesh. Shrikhande breathed his last on 21 April 2020. Some appreciations are in [88, 89, 94, 96, 97].

Let me conclude this chapter with a few lines about Shrikhande's family. He was married to Shakuntala, who was a school teacher. The couple had four children: Vijay (who passed away in early 1990), Asha (who lives with her husband in Michigan, USA), Mohan (an Emeritus Professor at Central Michigan University), and Anil (who has worked in several companies in the USA and served as the head of some major companies in Delhi).

As a tribute to this great researcher, the discrete mathematics community of India, through the Academy of Discrete Mathematics and Applications (ADMA), instituted the Professor Shrikhande Memorial Lecture, commencing at its twenty-first annual conference held in June 2025 at Cochin University of Science and Technology.

Rosemary Bailey, combinatorialist and statistician, writes:

On 24 and 25 October 2003, the American Mathematical Society held a meeting at the University of North Carolina in Chapel Hill. A Special Session on *Association Schemes: 1973–2003* ran for the whole meeting, organized by William J. Martin from Worcester Polytechnic Instistute and Dijen K. Ray-Chaudhuri from Ohio State University. Among the speakers were Sharadchandra Shrikhande's son Mohan S.

Shrikhande, who spoke on 'Delsarte polynomial and designs', and I, who spoke on 'Designs on association schemes'.

In the evening of 24 October, the speakers in this session all gathered for dinner at the Best Western. Mohan Shrikhande surprised and delighted everyone by bringing his father to the dinner. I was among those who managed to spend some time sitting next to him and talking about joint interests. During this conversation, Shrikhande senior told me that the proudest moment of his entire career was proving the Bose–Shrikhande–Parker theorem.

PART II

Graph Basics

2

Definitions

2.1 What Is a Graph?

'Graph theory' is a relatively recent addition to the mathematical canon. Some date its origins to the Swiss mathematician Leonhard Euler's (1708–83) work on the bridges of Königsberg (now Kaliningrad) in 1736. The city was at the junction of two rivers forming an island, and seven bridges connected the four regions of the city (Fig. 2.1). According to legend, the citizens asked Euler whether it was possible to take a walk, crossing each of the bridges just once and returning to the starting point. Euler showed that this was not possible.

(The picture shows the layout of the bridges in Euler's time. Since then, destruction in war and new construction have changed the layout. See [75] for a contemporary account.)

The problem is unchanged if we draw it in an abstract way as shown later in Fig. 2.2. Each of the four regions is represented by a dot, and each of the seven bridges by a line connecting two dots.

Such a diagram is a representation of a *graph*. However, the first usage of the term 'graph' was due to J. J. Sylvester [100] in 1878.

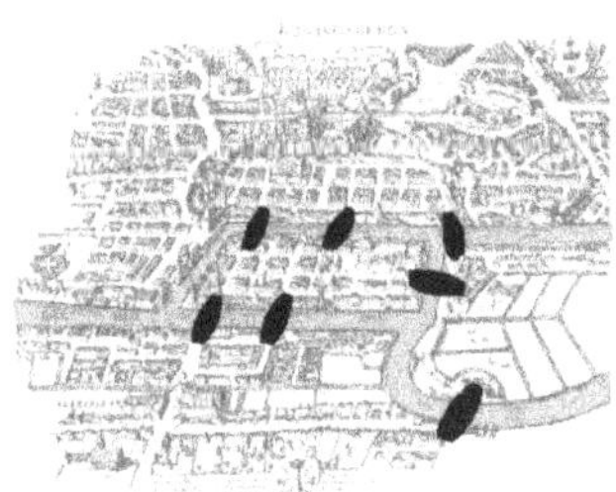

Figure 2.1 The bridges of Königsberg

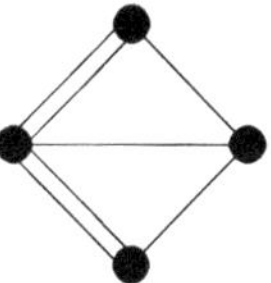

Figure 2.2 The bridges of Königsberg as a graph

Graphs play two main roles in mathematics and applications. First, as in Euler's example, they relate to *connectivity* properties of a structure. For the second role, suppose we have a network of radio transmitters, say for mobile phones. For reasons of basic physics, transmitters which are close together cannot both transmit on the same frequency. So the graph describing the network models *incompatibility*: if we think of frequencies as 'colours' assigned to the transmitters, then neighbouring transmitters must be given different colours. This application goes back to the famous *Four Colour Problem* from the nineteenth century, asking whether any map could be coloured with four colours so that countries sharing a border are given different colours.

We will refer several times to the history of graph theory. A good reference for this is the book by Biggs, Lloyd and Wilson [16].

2.2 Definitions and Basic Results

This section contains all the definitions and basic results that a beginning graph theory student might need. If you are familiar with these, skip to Section 2.3; you can refer back here when necessary. If you are learning graph theory, we have included exercises to illustrate the definitions.

Definition   A *graph* $G = (V, E)$ consists of a non-empty collection V of points called its *vertices*, and a set E of unordered pairs of distinct vertices called its *edges*. The unordered pair of vertices $\{u, v\} \in E$ are called the *end vertices* of the edge $e = \{u, v\}$. In that case, the vertex u is said to be *adjacent* to the vertex v. Two edges e and e' are said to be *incident* if they have a common end vertex. $|V|$ is called the *order* of G, denoted by n or $n(G)$; and $|E|$ is called the *size* of G, denoted by m or $m(G)$. A graph G is *trivial* or *empty* if it has no edges.

We sometimes abbreviate the edge $\{v, w\}$ to vw.

Note that this definition means that edges have no preferred direction, that no vertex is joined to itself, and that at most one edge joins a given pair of vertices. Graph theorists say that our graphs are *undirected*, *loopless* and *simple* (no *multiple edges*). In some parts of graph theory, these conditions are relaxed.

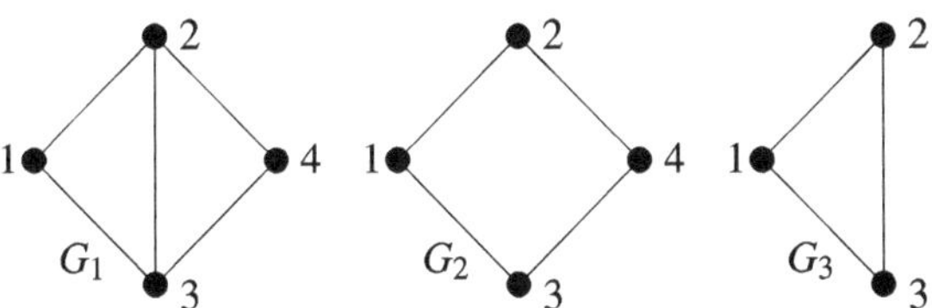

Figure 2.3 Spanning and induced subgraphs

(For example, the graph of the Königsberg bridges shown earlier is undirected and loopless but not simple.)

Exercise Show that there are eight graphs on a set of three vertices. How many 'essentially different' graphs are there? (By the term 'essentially different' we mean 'non-isomorphic graphs', which will be explained later.)

Definition A graph is *complete* if every pair of distinct vertices are the end vertices of an edge.

A complete graph on n vertices is denoted by K_n.

Exercise Prove that K_n has $\frac{n(n-1)}{2}$ edges.

Definition A graph $H = (V',E')$ is called a *subgraph* of G if $V' \subseteq V$ and $E' \subseteq E$. (This definition implies that, if $\{v,w\}$ is an edge in E', then both v and w belong to V'.) A subgraph H is a *spanning subgraph* if $V' = V$. H is called an *induced subgraph* if E' is the collection of all edges in G which has both its end vertices in V'. The induced subgraph with vertex set V' is denoted by $\langle V' \rangle$.

Figure 2.3 shows examples: G_2 is a spanning subgraph of G_1, while G_3 is an induced subgraph.

Definition A property P of a graph G is *vertex hereditary* if every induced subgraph of G has the property P. A graph H is a *forbidden subgraph* for a property P if any graph G that satisfies the property P cannot have H as an induced subgraph. A graph G is *H-free* if it does not have H as an induced subgraph.

Exercise Prove that, if a complete graph is a subgraph of a graph G, then it is an induced subgraph of G.

Definition The number of vertices adjacent to a vertex v is called the *degree* (or *valency*) of the vertex, denoted by $d(v)$. A vertex of degree 0 is called an *isolated vertex*, a vertex of degree one is called a *pendant vertex*, and a vertex of degree $n - 1$ is called a *universal vertex* (or *dominating vertex*).

Definition A graph G is *k-regular*, or *regular of degree k*, if $d(v) = k$ for every vertex $v \in V(G)$. It is *regular* if it is k-regular for some k. A spanning 1-regular graph is called a *1-factor* or *perfect matching*.

Exercise Prove that K_n is $(n - 1)$-regular.

Exercise How many of the three-vertex graphs found earlier are regular?

Exercise Prove the *Handshaking Lemma*: In any graph, the number of vertices of odd degree is even.

Definition A *path* on n vertices P_n is the graph with vertex set $\{v_1, v_2, \ldots, v_n\}$ and v_i is adjacent to v_{i+1} for $i = 1, 2, \ldots, n - 1$ are the only edges. If in addition v_n is adjacent to v_1, then it is called a *cycle* of length n, denoted C_n. A path from the vertex u to the vertex v is called a $u - v$ *path*. A graph G is *connected* if for every $u, v \in V$, there exists a $u - v$ path. If G is not connected, then it is *disconnected*. A maximal connected subgraph of G is called a *connected component* of G. A component of a graph G is *non-trivial* if it has at least one edge. A graph is *acyclic*, or a *forest*, if it does not contain cycles.

Definition A *tree* is a connected acyclic graph.

Exercise (a) Define a relation $\sim$ on the vertex set V of a graph G by the rule that $v \sim w$ if there is a path with end-points v and w (possibly of length 0, if $v = w$). Show that $\sim$ is an equivalence relation.
(b) Prove that the connected components of a graph G are the induced subgraphs on the equivalence classes of the relation $\sim$.

Thus, a graph is connected if it has a single connected component.

Exercise Show that a tree with more than one vertex has a vertex of degree 1. (Hint: if every vertex has a degree greater than 1, there is a cycle.) Hence, or otherwise, show that a tree with n vertices has $n - 1$ edges.

Exercise Show that a connected graph contains a *spanning tree* (a spanning subgraph which is a tree.)

Definition An *Eulerian path* in a graph G is a sequence

$$(v_0, v_1, \ldots, v_N)$$

of vertices of G such that, for each i with $1 \leq i \leq N$, the pair $\{v_{i-1}, v_i\}$ is an edge of G, and that every edge of G occurs exactly once in this list. It is an *Eulerian cycle* if $v_N = v_0$. A graph having an Eulerian cycle is called *Eulerian*.

Euler's solution to the bridges of Königsberg was Theorem 2.1.

Theorem 2.1 *A graph is Eulerian if and only if it is connected and every vertex has even degree. A connected graph in which just two vertices have odd degree has an Eulerian path.*

Informally, a graph has an Eulerian path if you can draw it without taking your pencil off the paper. Such puzzles are popularly known as 'Diagram tracing puzzles'. A search on the Internet will reveal large numbers of these. The graph describing the bridges of Königsberg has four vertices of odd degree, so it has no Eulerian cycle or path.

Exercise Draw a graph which has an Eulerian path. Indicate the path.

Exercise Does the complete graph K_6 have an Eulerian path? If not, find the minimum number of times a pencil has to be taken off the paper.

Exercise Prove Euler's theorem. Show also that if a connected graph has just two vertices of odd degree, then it has an Eulerian path, but any Eulerian path must start at one and end at the other.

Definition A *Hamiltonian cycle* in a graph is a spanning subgraph which is isomorphic to a cycle (a connected graph in which every vertex has degree 2). A graph having a Hamiltonian cycle is called *Hamiltonian*. Note that a Hamiltonian graph must be connected.

Hamiltonian graphs are named after the Irish mathematician William Rowan Hamilton (1805–1865). A recreational version of finding a Hamiltonian cycle is the 'icosian game', involving Hamiltonian cycles on the graph formed by the vertices and edges of the dodecahedron. Hamilton found this in connection with the construction of a non-commutative algebra [56], and turned it into a puzzle, which was marketed but was not a commercial success. For discussion see [16, p. 31].

So far we have discussed two important classes of graphs, the Eulerian graphs and the Hamiltonian graphs. Euler found a simple test (Theorem 2.1) to decide whether a graph is Eulerian. No such simple test is known for the Hamiltonian property; deciding whether a graph is Hamiltonian, and finding a Hamiltonian cycle if one exists, is a difficult problem in general.

A simple sufficient condition is given by *Dirac's theorem*:

Theorem 2.2 *If G is an n-vertex graph in which every vertex has degree at least $\frac{n}{2}$, then G is Hamiltonian.*

Definition   The set of all vertices adjacent to a vertex v is called the *open neighbourhood* of v, denoted by $N(v)$. The open neighbourhood of v together with the vertex v is called the *closed neighbourhood* of v, denoted by $N[v]$.

Definition   A *false twin* or *closed twin* of a vertex u is a vertex v which is adjacent precisely to all the vertices in $N[u] \setminus \{v\}$ (so that $N[u] = N[v]$). A *true twin* or *open twin* of a vertex u is a vertex v which is adjacent to precisely all the vertices in $N(u)$ (so that $N(u) = N(v)$). Note that false twins must be adjacent, while true twins must be non-adjacent.

Exercise   Show that every graph with three vertices has a pair of twins. What is the order of the smallest graph without twins? Find all such graphs of this minimum order.

Definition   A graph $G = (V,E)$ is *isomorphic* to a graph $H = (V',E')$ if there exists a bijection from V to V', which preserves adjacency. Such a bijection is called an *isomorphism*. If G is isomorphic to H, we write $G = H$.

An isomorphism from $G = (V,E)$ to itself (a permutation of V which preserves adjacency) is called an *automorphism* of G. The set of all automorphisms of G is closed under composition and forms a group, called the *automorphism group* of G.

Exercise   Illustrate two graphs which are isomorphic and two graphs which are not isomorphic.

When we referred to graphs being 'essentially different' earlier, we meant: How many equivalence classes (under the relation of isomorphism) of graphs on three vertices are there?

Exercise   Show that any graph-theoretic property, such as regularity or a particular forbidden subgraph, existence of twins or dominating vertices or connectedness, is shared by isomorphic graphs.

Exercise   Find the automorphism groups of C_4, P_n and $K_{m,n}$.

Exercise   Does there exist a graph whose automorphism group is S_n, the symmetric group on n elements?

After solving the previous exercise, the reader will realise that, for the complete graphs, any bijection defined on the vertex set is an automorphism. There are graphs for which the opposite extreme happens. Such graphs are called identity graphs.

Definition   A graph for which the only automorphism is the identity mapping is called an *identity graph*.

Exercise Show that the trees E_7 and E_8 in Fig. 8.3 in Section 8.4 are identity graphs.

What is the smallest number of vertices of an identity graph?

Exercise Show that isomorphic graphs have isomorphic automorphism groups. What are the automorphism groups of the graphs on three vertices?

Definition The graph G is the *disjoint union* of graphs G_1 and G_2 if $V(G)$ is the disjoint union of $V(G_1)$ and $V(G_2)$, and $E(G)$ is the disjoint union of $E(G_1)$ and $E(G_2)$. This extends to the disjoint union of any number of graphs in the obvious way.

Exercise Prove that any graph is the disjoint union of its connected components.

Exercise Which of the graphs on three vertices are connected?

Definition A graph G is *bipartite* if the vertex set can be partitioned into two non-empty sets U and U' such that every edge of G has one end vertex in U and the other in U'. Such a partition is called a *bipartition* of G. A bipartite graph in which each vertex of U is adjacent to every vertex of U' is called a *complete bipartite graph*. If $|U| = m$ and $U' = |n|$, then the complete bipartite graph is denoted by $K_{m,n}$. The complete bipartite graph $K_{1,n}$ is called a *star*.

Exercise Which of the graphs on three vertices are bipartite?

Exercise Show that a connected bipartite graph has a unique bipartition.

Exercise Show that a graph is bipartite if and only if it does not contain any odd cycle. Deduce, or prove otherwise, that a tree is bipartite.

Definition Let G be a graph. The *complement* of G, denoted by G^c, is the graph with vertex set same as that of V and any two vertices are adjacent in G^c if they are not adjacent in G. K_n^c is called *totally disconnected*. A graph G is called **self-complementary** if G is isomorphic to G^c. The path P_4 and the cycle C_5 are self-complementary graphs.

Exercise A couple of observations about complements:

- If a graph of order n is k-regular, then its complement is $(n - k - 1)$-regular.
- The order of a self-complementary graph is congruent to 0 or 1 (mod 4). Consequently, there are no self-complementary graphs of order 2, 3, 6, 7, 10, 11 and so on.
- A graph and its complement have the same automorphism group.

- If G is not connected, then its complement is connected with a diameter at most 2.

Definition A subset $I \subseteq V$ of vertices is said to be *independent* if no two vertices of I are adjacent. The maximum cardinality of an independent set is called the *independence number $\alpha(G)$*. A subset $K \subseteq V$ is called a *covering* of G if every edge of G is incident with at least one vertex of K. The number of vertices in a minimum covering is called the *covering number $\beta(G)$*.

Definition Recall that a subgraph H of G is *complete* if every pair of distinct vertices of H is adjacent. A complete subgraph is maximal if it is not properly contained in any other complete subgraph. A maximal complete subgraph is called a *clique*. The size of the largest clique in G is called the *clique number $\omega(G)$*.

Exercise Find the clique number of a complete graph, a complete bipartite graph, and cycles.

Definition The *intersection graph* of a family $\mathcal{F}$ of sets is the graph whose vertex set is the set $\mathcal{F}$, with two vertices adjacent if the corresponding sets have non-empty intersection. The intersection graph of all cliques of a graph G is called the *clique graph* of G, denoted by $K(G)$. If $K(G)$ is complete, then G is called *clique complete*.

In Fig. 2.4, G_1 is clique complete.

Exercise Every graph is an intersection graph. [Hint: Take the underlying set to be $X = V(G) \cup E(G()$, and the family $\mathcal{F} = \{S_v : v \in V(G)\}$ of subsets of X, where $S_v = \{v\} \cup E(v)$ and $E(v)$ is the set of edges containing the vertex v.]

A graph H is a clique graph if there exists a graph G such that $H = K(G)$. Not every graph is a clique graph:

Exercise Find examples of graphs which are not clique graphs.

Exercise Find the clique graphs of paths, cycles, complete bipartite graphs and complete graphs.

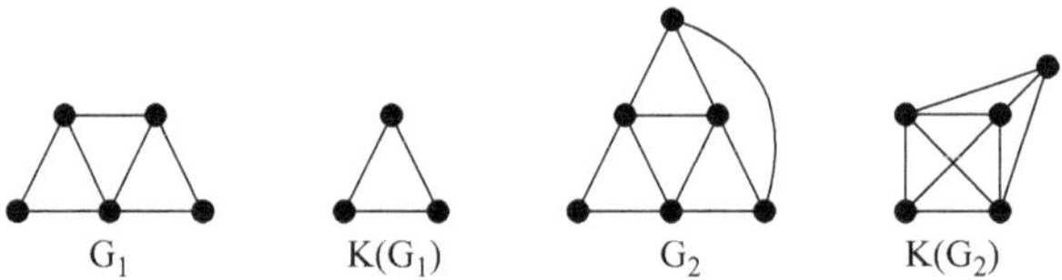

Figure 2.4 Example of a clique graph of a graph

A function whose domain and codomain are collections of some (or all) graphs is called a graph operator. At the end of the above exercise, the reader can see that there are graphs for which $K(G) = G$. Such graphs are called fixed graphs with respect to the graph operator which maps a graph to its clique graph. A branch of Graph Theory, called *Graph Dynamics*, deals with properties of the sequence $S = (G, \phi(G), \phi(\phi(G)), \ldots$ obtained by iterating the operator ϕ applied to G. In particular, in the literature it is said that

- the graph G *converges* under ϕ if $\{\phi^n(G) : n \in \mathbb{N}\}$ is finite and *diverges* otherwise;
- the graph is *periodic* under ϕ if $\phi^k(G) \cong G$ for some k, and in particular is *fixed* if this holds with $k = 1$;
- the graph G is *mortal* if $\phi^k(G)$ is empty for some k.

We note that a graph converges in this sense if the sequence is ultimately periodic (that is, $\phi^k(G) = \phi^l(G)$ for some $k < l$. As an example, paths are mortal with respect to the operator L mapping a graph to its line graph. Interested readers may refer to [81].

Definition A collection of objects $\mathscr{E}$ satisfies the *Helly property* if for any sub collection $\mathscr{E}' \subseteq \mathscr{E}$, the elements of $\mathscr{E}'$ pairwise intersect, then $\bigcap_{e \in \mathscr{E}'} e \neq \varnothing$. If the cliques of a graph G satisfy the Helly property, then we say that G is *clique-Helly*. If G and all its induced subgraphs are clique-Helly, then G is *hereditary clique-Helly*.

In Fig. 2.4, G_1 is clique-Helly, whereas G_2 is not.

Exercise Prove that all graphs of order less than six are clique-Helly.

Definition An assignment of colours to the vertices of a graph is called a *vertex colouring*. If no two adjacent vertices receive the same colour, the colouring is called a *proper vertex colouring*. The minimum number of colours required for a proper vertex colouring of a graph G is called its *chromatic number*, denoted by $\chi(G)$.

Exercise Find the chromatic number of K_n, P_n and C_n for $n \in \mathbb{N}$.

Exercise Prove that the chromatic number of G is at least as large as its clique number.

Exercise Prove that the chromatic number of G is at most two if and only if it is bipartite.

Definition A graph whose clique number and chromatic number are equal is called *weakly perfect*. A graph G is called *perfect* if all of its induced subgraphs (including G itself) are weakly perfect.

Exercise Show that a bipartite graph is perfect.

The celebrated (weak) Perfect Graph Theorem due to L. Lovász [72] states that the complement of a perfect graph is also perfect. The renowned French graph theorist Claude Berge had conjectured this in 1961 and had further conjectured that perfect graphs are precisely those that have neither odd holes (odd-length induced cycles of length at least five) nor odd antiholes (complements of odd holes) as induced subgraphs. Maria Chudnovsky, Neil Robertson, Paul Seymour and Robin Thomas published a proof of this conjecture [38] in 2006.

Definition The *distance* between two vertices u and v of a connected graph G, denoted by $d_G(u,v)$ or $d(u,v)$ is the length of a shortest path from u to v. The *eccentricity* of a vertex $e(v) = max\{d(u,v) : v \in V(G)\}$. The *radius* of a graph $r(G)$ is the minimum of the eccentricities of its vertices and the *diameter* of a graph $d(G)$ is the maximum of the eccentricities of its vertices.

Exercise For any connected graph G, prove that $r(G) \leq d(G) \leq 2r(G)$.

Exercise Find the radius and diameter of the graph shown in Fig. 2.5. What do you observe?

This graph is the celebrated *Petersen graph*. We will say more about it Chapter 3.

Definition A graph G for which $r(G) = d(G)$ is called a *self-centred graph*.

Exercise Is it true that for every n, there exists a self-centred graph of order n? Justify.

Exercise Let G be a connected graph. Show that

(a) $d_G(u,v) \geq 0$. with equality if and only if $u = v$.

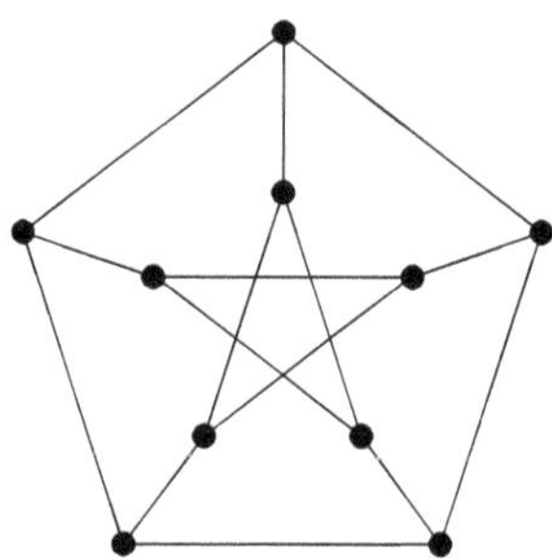

Figure 2.5 The Petersen graph

(b) $d_G(u,v) = d_G(v,u)$.

(c) $d_G(u,v) + d_G(v,w) \geq d_G(u,w)$, for any three vertices u,v,w.

A function d satisfying these three conditions is called a *metric*. The function d_G is the *graph metric* induced by G. The third condition is known as the *triangle inequality*.

Exercise Let G be a connected regular graph with diameter d and degree k. Show that

$$|VG)| \leq 1 + k + k(k-1) + \cdots + k(k-1)^{d-1}.$$

(If equality holds, then G is a *Moore graph*. We will discuss the existence of Moore graphs later.)

Exercise Show that a connected bipartite graph with diameter 2 is complete bipartite.

Definition The *line graph* of a graph G, denoted by $L(G)$, has the edges of G as its vertices, and any two vertices are adjacent in $L(G)$ if the corresponding edges in G are incident. The *iterated line graphs* of G are defined as $L^k(G) = L(L^{k-1}(G))$ for $k > 1$.

A graph G is a line graph if there exists a graph H such that $L(H) = G$.

The line graph of G is the intersection graph (see Section 2.2) of its edge set. But although, as we saw there, every graph is an intersection graph, not every graph is a line graph. As we will see later, the adjacency matrix of a line graph has smallest eigenvalue -2 or greater, and this property characterises a class slightly larger than the class of line graphs.

Exercise (a) Draw $L^k(G)$ for $k = 1,2,3,4$ of the following graph G. What do you observe?

(This is the F-graph (Fig. 2.6). Later we will meet it again under the name D_5.)

(b) Draw $L^k(P_n)$, for $k \geq 1$. What did you observe?

(c) Prove that if G is a k-regular graph, then $L(G)$ is $(2k-2)$-regular.

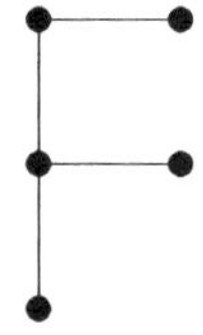

Figure 2.6 The F-graph

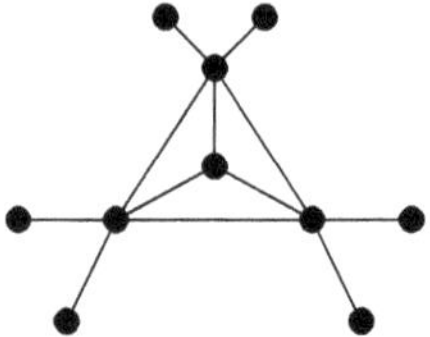

Figure 2.7 A graph G with $\gamma(G) = 3$, $\gamma_g(G) = 4$ and $\gamma_i(G) = 5$

(d) Draw the line graphs of K_3 and $K_{1,3}$. What did you observe?

(e) Prove that $K_{1,3}$ is not a line graph.

Definition A set $S \subseteq V$ of vertices in a graph G is called a *dominating set* if every vertex $v \in V$ is either an element of S or is adjacent to an element of S. (Thus, a universal or dominating vertex as defined earlier is just a vertex constituting a dominating set of size 1.) A dominating set S is *minimal dominating* if no proper subset of S is a dominating set. The *domination number $\gamma(G)$* of a graph G is the minimum cardinality of a dominating set in G. A set $S \subseteq V$ of vertices in a graph G is called a *global dominating set* if it dominates both G and G^c. The minimum cardinality of a global dominating set is called the *global domination number $\gamma_{gcd}(G)$*. A set $S \subseteq V$ of vertices in a graph G is called an *independent dominating set* if S is independent and S dominates G. The minimum cardinality of an independent dominating set is called the *independent domination number $\gamma_i(G)$*.

For the graph G shown in Fig. 2.7, the three vertices of degree 5 form a dominating set; these vertices together with a pendent vertex form a global dominating set and any two pairs of pendent vertices with same root together with the root vertex of the remaining pair of pendent vertices gives you an independent dominating set.

The 'Chessboard Problem' [105], which has a long history [7], has motivated many concepts in graph theory. In particular, the 'Queen's Domination' problem led to domination theory in graphs. The relevant question was 'How many queens are needed so that every square is attacked or occupied by one of the queens?'. This problem can be asked for other types of chess pieces and is known as the domination (or covering) problem.

Exercise Prove that, on an 8×8 chessboard, the minimum numbers of dominating kings, queens, rooks, bishops and knights are, respectively, 9, 5, 8, 8, 12.

Definition A graph that can be reduced to an edgeless graph by taking complements within components is called a *cograph*.

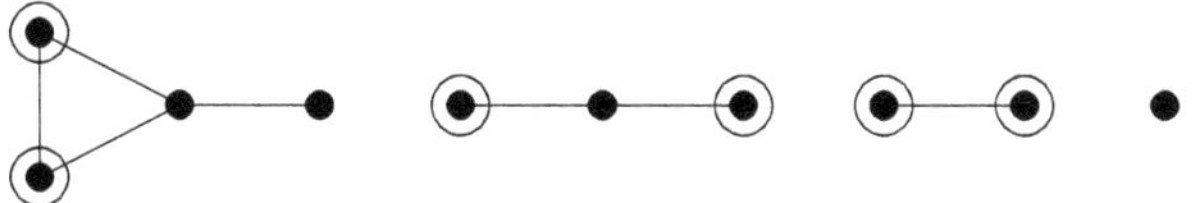

Figure 2.8 An example of twin reduction

For example, any graph of order less than or equal to four, except P_4, is a cograph. The complete bipartite graphs and complete graphs are also examples of cographs.

Cographs arise in many different areas; as a result they have been rediscovered several times and given different names (including *complement-reducible graphs* and *hereditary Dacey graphs*, see [40, 99]). They have several different characterizations, which we summarise next. First we define the process of *twin reduction* on a graph: choose a pair of twins; identify them (or, equivalently, delete one); then repeat the process until no twins remain. Identifying twins can create new twins. Here is an example of twin reduction (Fig. 2.8); the twins at each stage are circled.

There is some choice in the pair of twins at each step. However, we have:

Proposition 2.3 *The result of twin reduction of a graph is independent of the reduction, in the sense that any two reduced graphs are isomorphic.*

Here are several characterizations of cographs.

Theorem 2.4 *The following are equivalent for a finite graph G:*

(a) G is a cograph;
(b) G can be built from 1-vertex graphs by the operations of disjoint union and complementation;
(c) the result of twin reduction of G is a 1-vertex graph;
(d) G contains no induced subgraph isomorphic to the path P_4.

Definition A *plane representation* of a graph G is an isomorphic copy of G in which the vertices are points of the Euclidean plane, edges are continuous curves whose endpoints are their vertices, and in which edges intersect only at their vertices. A graph which admits a plane representation is called a *planar graph*; otherwise it is a *non-planar graph*.

There is a famous characterization of planar graphs by Kuratowski [69]. We invite readers to prove that K_5 and $K_{3,3}$ are non-planar.

Definition Let G be a graph. The process of *subdividing edges* of G consists of replacing the edges of G by paths of positive length so that two of these paths

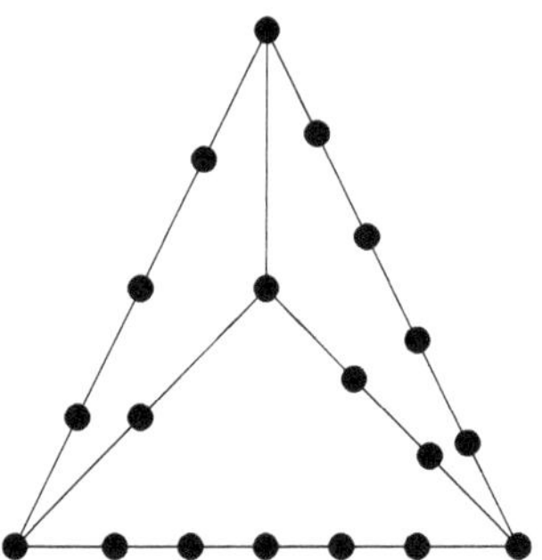

Figure 2.9 A subdivision of a graph

intersect only at their end vertices: this is formed by repeatedly replacing an edge $e = \{v, w\}$ by one new vertex x and two edges $\{v, x\}$ and $\{x, w\}$. (Informally we insert a new vertex into the edge.) Clearly this operation does not affect embeddability in a surface. Figure 2.9 shows a subdivision of a graph.

Theorem 2.5 (Kuratowski) *A graph G is embeddable in the plane if and only if it does not contain a subdivision of the complete graph K_5 or the complete bipartite graph $K_{3,3}$ as a subgraph.*

Definition The *union* of two graphs G and H, denoted by $G \cup H$, is the graph with vertex set $V(G) \cup V(H)$ and edge set $E(G) \cup E(H)$. If the vertex sets of G and H are disjoint then we call this the *disjoint union*.

Definition The *join* of two graphs G and H denoted by $G \vee H$ is the graph with vertex set $V(G) \cup V(H)$ and $E(G \vee H) = E(G) \cup E(H) \cup \{uv : u \in V(G)$ and $v \in V(H)\}$.

Definition The *tensor product* (also called the *categorical product*) of two graphs G and H, denoted by $G \times H$, is the graph with $V(G \times H) = \{u, v) : u \in V(G_1)$ and $v \in V(G_2)\}$ and any two vertices (u_1, v_1) and (u_2, v_2) are adjacent if $u_1 u_2 \in E(G_1)$ and $v_1 v_2 \in E(G_2)$.

Definition The *Cartesian product* of two graphs G and H, denoted by $G \square H$, is the graph with $V(G \square H) = \{(u, v) : u \in V(G_1)$ and $v \in V(G_2)\}$ and any two vertices (u_1, v_1) and (u_2, v_2) are adjacent if one of the following holds:

- $u_1 = u_2$ and $v_1 v_2 \in E(G_2)$;
- $u_1 u_2 \in E(G_1)$ and $v_1 = v_2$.

Definition The *strong product* of two graphs G and H, denoted by $G \boxtimes H$, is the graph with $V(G \boxtimes H) = \{u, v\} : u \in V(G_1)$ and $v \in V(G_2)\}$ and any two vertices (u_1, v_1) and (u_2, v_2) are adjacent if one of the following holds:

(a) $u_1 = u_2$ and $v_1 v_2 \in E(G_2)$;
(b) $u_1 u_2 \in E(G_1)$ and $v_1 = v_2$;
(c) $u_1 u_2 \in E(G_1)$ and $v_1 v_2 \in E(G_2)$.

Note that it is the union of the tensor and Cartesian products (which have the same vertex sets but disjoint edge sets). The notation $G \boxtimes H$ is chosen to suggest this. In fact, the symbol used for each of the three products represents the corresponding product of two copies of K_2.

Exercise Prove that the graphs $K_3 \,\square\, K_3$ and $K_3 \times K_3$ are isomorphic. Show also that these graphs are complements of each other (so that each of them is self-complementary).

This exercise also indicates that it can happen that a graph product can be self-complementary, without its factors being self-complementary.

Exercise Show that $K_m \,\square\, K_n$ is isomorphic to the line graph of the complete bipartite graph $K_{m,n}$.

These products can be extended in the obvious way to more than two factors. For example, consider the Cartesian product of n copies of the graph K_2 with two vertices and a single edge. The vertex set can be identified with the set of all n-tuples of 0s and 1s; two vertices are joined if they agree in all coordinates except one. This graph is the n-dimensional cube, denoted by Q_n. (To see the reason for the name, consider an ordinary unit cube in three-dimensional Euclidean space. Its vertices and edges are exactly as described.)

Remark Another graph product not mentioned here is the *lexicographic product*. The book [57] by Hammack, Imrich and Klavžar is recommended for further information on graph products.

Definition A graphical invariant σ is *supermultiplicative* with respect to a graph product $\circ$, if given any two graphs G and H, $\sigma(G \circ H) \geq \sigma(G)\sigma(H)$ and *submultiplicative* if $\sigma(G \circ H) \leq \sigma(G)\sigma(H)$. A class $\mathscr{C}$ is called a *universal multiplicative* class for σ on $\circ$ if for every graph H, $\sigma(G \circ H) = \sigma(G)\sigma(H)$ whenever $G \in \mathscr{C}$.

A famous conjecture due to V. J. Vizing [103] states that the domination number is supermultiplicative with respect to Cartesian product:

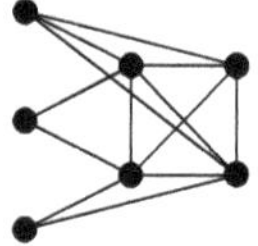

Figure 2.10 Example of a split graph

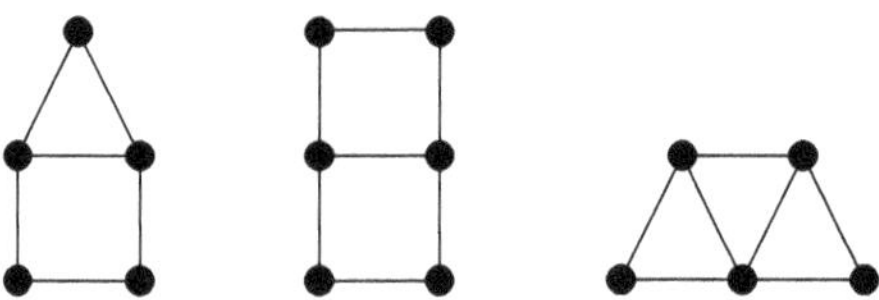

Figure 2.11 House, Domino and Gem graphs

Vizing's Conjecture: For any two graphs G and H, $\gamma(G \,\square\, H) \geq \gamma(G)\gamma(H)$.

Exercise Show that clique number is submultiplicative for the tensor product.

Definition A graph G whose vertex set can be partitioned into an independent set and a clique is called a *split graph*.

Figure 2.10 gives an example of a split graph.

Definition A graph G is a *threshold graph* if it can be obtained from K_1 by recursively adding isolated vertices and universal vertices.

There is an alternative characterization that explains the name. A graph G is a threshold graph if there are real numbers $s(v)$ associated with the vertices v and a real number t called the threshold, so that vw is an edge if and only if $s(v) + s(w) \geq t$.

The two classes just defined also have characterizations by forbidden induced subgraphs.

Definition A graph G is a *distance hereditary graph* if, for every connected induced subgraph H of G, we have $d_H(u,v) = d_G(u,v)$.

Distance hereditary graphs admit a forbidden subgraph characterization:

Proposition 2.6 *A graph G is distance hereditary if and only if it does not contain an induced house, hole, domino or gem, where a hole is a cycle of length greater than five and the other graphs are shown in Fig. 2.11.*

Cographs form a subclass of distance hereditary graphs. A graph G is a cograph if and only if it is the disjoint union of distance hereditary graphs of diameter at most two. In [8], Bandelt and Mulder had made an intensive study of distance hereditary graphs, which plays a significant role in 'convexity of graphs'.

2.3 Algebraic Graph Theory

How do we describe a graph? If it is small, we can draw it. But suppose it is too large to draw, and has to be input to a computer system. One way to do this is to use the *adjacency matrix* of the graph.

Suppose that the vertex set $V(G)$ of G has n elements $v_1, v_2, \ldots, v_n$. Then we can represent the graph by an $n \times n$ matrix, called the *adjacency matrix* of G, as follows: the entry in row i and column j is given by

$$A_{ij} = \begin{cases} 1 & \text{if } v_i \text{ and } v_j \text{ are adjacent,} \\ 0 & \text{otherwise.} \end{cases}$$

If the graph is undirected, then A is a real symmetric matrix, a member of the nicest class of matrices in terms of algebraic theory. There is an orthonormal basis for $\mathbb{R}^n$ consisting of *eigenvectors* $x_1, \ldots, x_n$ of $A(G)$; that is, to each eigenvector, there is a corresponding *eigenvalue* θ_i so that

$$A(G)x_i = \theta_i x_i.$$

So, if we use $x_1, x_2, \ldots, x_n$ as a new basis for the vector space $A(G)$, the matrix becomes diagonal.

Remark The word 'eigenvalue' is a mixture of English and German; they are sometimes called characteristic values or proper values. The term 'latent roots' is also used. This dates from the early days of photography, when the image on the film was latent until a certain developing process was applied. In much the same way, the latent roots of a matrix are hidden in the matrix, and are revealed by writing down the *characteristic polynomial* $c_A(x) = \det(xI - A)$ of the matrix A and solving the polynomial equation $c_A(x) = 0$.

Often we will shorten 'eigenvalue of the adjacency matrix of G' to 'eigenvalue of G'. Note that isomorphic graphs have the same eigenvalues with the same multiplicities.

Exercise Let A and B be the adjacency matrices of isomorphic graphs. Show that $B = P^{-1}AP$, where P is a *permutation matrix*, a matrix with just one nonzero element (equal to 1) in each row and each column. Deduce that A and B have the same eigenvalues.

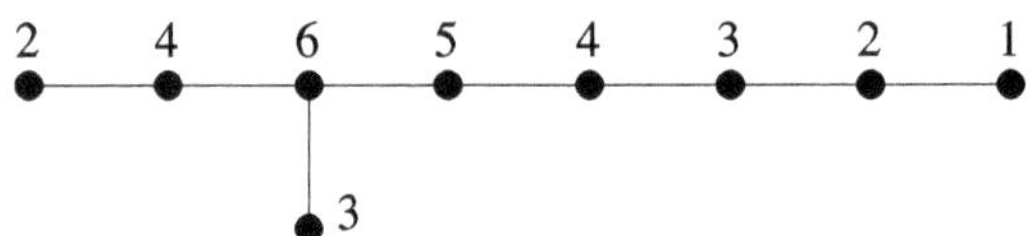

Figure 2.12 An eigenvector of E_8

In the case of graphs, there is a simple interpretation of eigenvalues and eigenvectors, which we will meet again. Consider the following of Fig. 2.12. We have a graph, with numbers at the vertices, with the special property that the sum of the labels on the neighbours of any vertex is always twice the label of the vertex.

This translates into the equation

$$A(G)x = 2x,$$

where $A(G)$ is the adjacency matrix of the graph, and x the vector whose ith entry is the label of the ith vertex. So the labelling gives us an eigenvector of the graph with eigenvalue 2.

Any eigenvalue and eigenvector of a graph have a similar interpretation. But there is something special about the situation in the diagram: all the labels are positive. This illustrates the *Perron–Frobenius theorem*, at least in the special case of graphs, stated here:

Theorem 2.7 *Let G be a connected graph. Then, up to scalar multiples, there is a unique eigenvector of $A(G)$ all of whose entries are positive. The corresponding eigenvalue is the greatest eigenvalue of $A(G)$.*

Note that, if a graph is regular with valency k, then the all-1 vector is an eigenvector with eigenvalue k; this is the Perron–Frobenius eigenvector.

The greatest and least eigenvalues of a graph G have special importance. Note that if a graph is not null, then its greatest eigenvalue is positive; but the sum of the eigenvalues is equal to the trace of $A(G)$, which is zero (since all the diagonal elements of $A(G)$ are zero as there are no loops); so the least eigenvalue is negative. If G is k-regular, $|\lambda| \leq k$; that is, the eigenvalue lies in the interval $[-k,k]$. However, there are certain graph classes and certain intervals such that the eigenvalues of the graph classes will not lie in these intervals. As an example, if G is a cograph, then the interval $(-1,0)$ contains no eigenvalue of G. Moreover, the multiplicities of -1 and 0 give information about twin vertices in G [78].

The eigenvalues carry important information about the graph. For an example, we can recognise whether a bipartite graph is bipartite or regular:

Theorem 2.8 *Let G be a connected graph with largest eigenvalue λ. Then the following conditions are equivalent:*

(a) $-\lambda$ is an eigenvalue of G;
(b) the spectrum of G is symmetric about the origin;
(c) G is bipartite.

We will sketch the main ideas of the proof. If G is bipartite, then its adjacency matrix has the form

$$A(G) = \begin{pmatrix} O & B \\ B^\top & O \end{pmatrix}.$$

If $[v, w]$ is an eigenvector with eigenvalue μ, then $vB = \mu w$ and $wB^\top = \mu v$; a short calculation shows that $[v, -w]$ is an eigenvector with eigenvalue $-\mu$. So the spectrum is symmetric.

Conversely, note that if $-\lambda$ is an eigenvalue, then the positions of the positive and negative entries of a corresponding eigenvector give a bipartition of G.

Theorem 2.9 *Let G be a graph on n vertices with eigenvalues $\lambda_1, \ldots, \lambda_n$, where λ_1 is the greatest. Then G is regular if and only if*

$$\sum_{i=1}^{n} \lambda_i^2 = n\lambda_1.$$

For another application, we state *Hoffman's bound*. We will come across Alan Hoffman again in this book, but he did not publish this bound; we refer to Haemers [55] for the history of this.

Theorem 2.10 (Hoffman's ratio bound) *Let G be a regular graph on n vertices with valency k, and let λ be its least eigenvalue. Then the independence number $\alpha(G)$ of G satisfies*

$$\alpha(G) \le \frac{-n\lambda}{k - \lambda}.$$

Note that this is a bound, which may or may not be met. The two stars of our story, the line graph of $K_{4,4}$ and the Shrikhande graph, both have valency $k = 6$ and least eigenvalue $\lambda = -2$. So the bound for the independence number is $\frac{16 \times 2}{6+2} = 4$. Both graphs have independence number 4, meeting the bound: in $L(K_{4,4})$, take a matching in $K_{4,4}$, while a set in the Shrikhande graph can be found by trial and error. But the complementary graphs both have valency 9 and smallest eigenvalue -3, so Hoffman's bound gives $\frac{16 \times 3}{9+3} = 4$. An independent set in the complement is a clique in the original graph, and $L(K_{4,4})$ has cliques of size 4 (the four edges through a vertex in $K_{4,4}$) whereas the largest clique in the Shrikhande graph has size 3.

Definition A graph G is *integral* if all its eigenvalues are integers.

The complete graph and the Petersen graph are examples of integral graphs, and we will come across several more, including the Shrikhande graph. A detailed survey on integral graphs is available in [6].

2.4 Laplacian Matrix

The Laplacian matrix of a graph G is a variant of the adjacency matrix which is useful in some contexts.

Definition Let G be a graph with vertex set $\{v_1, \ldots, v_n\}$. The *Laplacian matrix* of G is the $n \times n$ matrix $\mathscr{L}(G)$ with (i,j) entry

$$
\mathscr{L}(G)_{ij} = \begin{cases} d(v_i) & \text{if } j = i, \\ -1 & \text{if } \{v_i, v_j\} \in E(G), \\ 0 & \text{otherwise,} \end{cases}
$$

where $d(v_i)$ is the degree of v_i.

In other words, $\mathscr{L}(G) = D(G) - A(G)$, where $D(G)$ is the diagonal matrix of vertex degrees and $A(G)$ the adjacency matrix.

Note that, if G is regular, then $\mathscr{L}(G)$ is a linear combination of I and $A(G)$, so its spectrum can be computed from the spectrum of $A(G)$.

Note also that all row and column sums of $\mathscr{L}(G)$ are zero, so its determinant is zero. It can be shown that G is connected if and only if 0 is a simple eigenvalue of $\mathscr{L}(G)$.

Theorem 2.11 *(a)* $\mathscr{L}(G)$ *is positive semidefinite.*
(b) If G is connected, then the following three quantities are equal: the number of spanning trees of G; the absolute value of any cofactor of $\mathscr{L}(G)$; and the product of the nonzero eigenvalues of $\mathscr{L}(G)$ divided by n.

This is Kirchhoff's celebrated *Matrix-Tree Theorem*. The (i,j) cofactor of a matrix is obtained by deleting row i and column j, calculating the determinant, and multiplying by $(-1)^{i+j}$.

The smallest non-zero Laplacian eigenvalue of a graph is a very important parameter of a graph, being related to expansion properties, rate of convergence of the random walk on the graph and (in statistics) optimality properties of a design having the graph as its concurrence graph. We refer to [2] for discussion and proof of the Matrix-Tree Theorem, and the following application of the smallest positive Laplacian eigenvalue.

Definition The *isoperimetric number* $\iota(G)$ of a connected graph G is defined as

$$\iota(G) = \min\left\{ \frac{|\partial S|}{|S|} : S \subseteq V(G), 0 < |S| \leq |V|/2 \right\},$$

where ∂S is the set of edges with one end in S and the other in $V(G) \setminus S$.

Theorem 2.12 *Let G be a connected graph whose smallest non-zero Laplacian eigenvalue is μ. Then $\iota(G) \geq \mu/2$.*

2.5 Graph Polynomials

A colouring of the vertices of a graph such that no two adjacent vertices are of the same colour is called a (proper) vertex colouring. So, a k-colouring is a mapping $f : V(G) \to \{1, 2, 3, \ldots, k\}$ such that if u and v are adjacent in G, then $f(u) \neq f(v)$. The minimum number of colours required for such a colouring is called the *chromatic number* of G, denoted by $\chi(G)$. The term 'colouring' is for historical reasons. While trying to colour a map of the counties of England, Francis Guthrie (1831–1899), noted that four colours were sufficient to colour the map so that no regions sharing a common border received the same colour. This observation later on became the celebrated 'four colour conjecture' [109].

(Frederick Guthrie, a brother of Francis Guthrie, passed on the question to his mathematics teacher Augustus De Morgan at University College, who mentioned it in a letter to William Hamilton in 1852. Arthur Cayley raised the problem at a meeting of the London Mathematical Society in 1879. The same year, Alfred Kempe published a paper that claimed to establish the result, and for a decade the four colour problem was considered solved. Surprisingly, Kempe had a unique method of 'proof' and for this accomplishment, he was elected a Fellow of the Royal Society and later President of the London Mathematical Society; but his proof was later found to be defective. Despite the fact that this conjecture was tried by many pioneers in graph theory, it defied a solution till 1976, when it was proved by Kenneth Appel and Wolfgang Haken, assisted in some algorithmic work by John A. Koch.)

George Birkhoff was among those who studied the four-colour conjecture. He and Lewis [17] introduced the function $P(G,k)$ which gives the number of proper colourings of G with k colours, for each positive integer k. Note that the chromatic number $\chi(G)$ of G is the smallest positive integer k such that $P(G,k) > 0$. $P(G,k)$ is the *chromatic polynomial* of G. It was introduced by Birkhoff and Lewis [17].

Definition A polynomial is *chromial* if it is the chromatic polynomial of some graph.

It is not true that every polynomial is chromial. The problem of characterizing them is not solved.

A tree of order n has chromatic polynomial $x(x-1)^{n-1}$, and indeed a graph with this chromatic polynomial must be a tree. (The fact that 0 is a simple root shows that it is connected; since the coefficient of x^{n-1} is $-(n-1)$, it has n vertices and $n-1$ edges.)

Birkhoff and Lewis showed the following result:

Theorem 2.13 *The function $P(G,k)$ is a polynomial in k; its degree is the number of vertices of G, and its leading coefficient is 1.*

Proof The shortest proof of this theorem uses the *Principle of Inclusion and Exclusion*: Let X be a set, and A_i a subset of X for $i \in I$. For $J \subseteq I$, let A_J be the set of elements of X lying in all the sets A_j for $j \in J$ (so that $A_\emptyset = X$). Then the number of elements of X which lie in none of the sets A_i is

$$\sum_{J \subseteq I} (-1)^{|J|} \cdot |A_J|.$$

Now let X consist of all the assignments of colours $c_1, \ldots, c_k$ to vertices of G: that is, all the colourings, proper and improper. For each edge $e = \{v,w\}$ of G, let A_e be the set of colourings with $c(v) = c(w)$. For any subset J of the set E of edges, let $c(J)$ be the number of connected components of the graph $(V(G),J)$ with edge set J. Note that, if a colouring of the vertices lies in A_e for all $e \in J$, then all the vertices in the same component of the graph $(V(G),J)$ have the same colour, and so there are $k^{c(J)}$ such colourings. So the number of proper colourings is

$$P(G,k) = \sum_{J \subseteq E} (-1)^{|J|} k^{\kappa(J)}.$$

The right-hand side is a signed sum of powers of k; the highest power is k^n, coming from the term with $J = \emptyset$.					$\square$

The roots of a chromatic polynomial are algebraic integers, since the polynomial is monic. It was conjectured by Cameron and Morgan [36] that *for every real algebraic integer α, there exists an integer n such that $\alpha + n$ is the root of a chromatic polynomial.* This has been proved for algebraic integers of degrees 2 and 3.

Two variants of the chromatic polynomial are

(a) the *Tutte polynomial*, or *rank generating function*, a two-variable polynomial which specializes to the chromatic polynomial. This was introduced (in slightly different forms) by Tutte [101] and Whitney [106].
(b) the *orbital chromatic polynomial*, $OP(G,\Gamma,k)$, where G is a graph and Γ a subgroup of the automorphism group of k, whose value for a given k is the number of orbits of Γ on proper colourings with k colours [34].

Another important graph polynomial is the *characteristic polynomial* of its adjacency matrix, the polynomial $\det(xI - A)$ where A is the adjacency matrix, which we met in Section 2.3.

There are several more graph polynomials. Here are two (but we will not discuss their properties further).

Definition The *independence polynomial* of a graph G is the polynomial in which the coefficient of x^k is the number of independent sets of size k in G. Its degree is the independence number $\alpha(G)$ of G.

The *matching polynomial* of G is the polynomial in which the coefficient of x^k is the number of matchings of size k in G. Its degree is the size of a maximal matching in G.

It is exciting to note that they also find applications in Mathematical Chemistry [54].

2.6 Enter the Shrikhande Graph

We shall now give a simple description of the Shrikhande graph, the main topic of our book. Let $\mathbb{Z}_4$ denote the integers mod 4. Now the vertex set of the graph is the set of all triples (x,y,z) with $x,y,z \in \mathbb{Z}_4$ and $x + y + z = 0$ (where the addition is modulo 4). Any two of x,y,z determine the third uniquely, so there are 16 vertices. We join two vertices (x,y,z) and (x',y',z') if $x' = x$, $y' = y + 1$, $z' = z - 1$, or any similar condition obtained by permuting the three coordinates x,y,z. Since there are six permutations, the valency of the graph is 6. We invite readers to use this description to show that any two vertices have two common neighbours.

This definition makes it simple to construct the Shrikhande graph in a suitable computer system. For example, in GAP 4 [47], using the package GRAPE [98], the following code produces the graph:

```
a:=(1,2,3,4); b:=(5,6,7,8);
G:=Group([a,b]); S:=[a,a^-1,b,b^-1,a*b^-1,b*a^-1];
SG:=CayleyGraph(G,S);
```

(See Section 4.3 for Cayley graphs. We have ignored the third coordinate of $\mathbb{Z}_4^3$, which is determined by the other two using the condition $x + y + z = 0$.)

Using the computer system suggested previously, one can use the GRAPE command `GlobalParameters(SG)` to verify the degree and the numbers of common neighbours of two vertices.

We will see several further descriptions of the graph as we proceed.

Exercises

2.1 Show that the vertices of a tree can be enumerated as $v_1, v_2, \ldots, v_n$ such that, for all $i > 1$, the vertex v_i has a unique neighbour in the set $\{v_1, \ldots, v_{i-1}\}$.

2.2 Show that any two of the following properties of a graph with n vertices imply the third:

- G is connected;
- G contains no cycle;
- G has $m - 1$ edges.

2.3 Let G be a self-complementary graph on the vertex set V. This means that G is isomorphic to its complement; this isomorphism is a permutation of V which carries edges to non-edges and vice versa. Prove that this permutation has at most one fixed point, and that the length of any other cycle is divisible by 4.

2.4 If G is connected but not a tree, then the girth of G is at most $2d + 1$, where d is the diameter of G.

Suppose that G has n vertices, and is connected with diameter d and regular of degree k. Show that G has girth $2d + 1$ if and only if

$$n = 1 + k + k(k - 1) + \cdots + k(k - 1)^{d-1}.$$

A graph satisfying these conditions is called a *Moore graph*.

2.5 Prove Proposition 2.3 and Theorem 2.4.

2.6 Let S be a set of vertices in a graph G, such that any two vertices in S are twins. Prove that all such pairs are the same kind of twins (all closed, or all open). Hence prove that any permutation of S, fixing all other vertices of G, is an automorphism of G.

2.7 Show that if a graph G has m edges, then

$$\chi(G) \le \frac{1}{2} + \sqrt{2m + \frac{1}{4}}.$$

2.8 Show that a graph G is bipartite if and only if its chromatic number is at most 2.

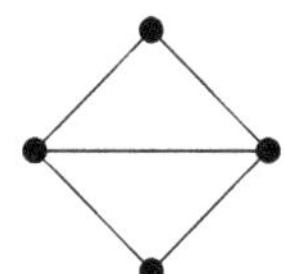

Figure 2.13 A graph

2.9 Find the chromatic polynomial $P(G,k)$ of the graph G in Fig. 2.13, and verify that $\chi(G) = 3$.

Find the automorphism group of the graph, and calculate the number of orbits on proper three-colourings.

2.10 Using the description of the Shrikhande graph in Section 2.6, show:

(a) it is strongly regular, with the parameters claimed;
(b) the induced subgraph on the neighbourhood of a vertex is a six-cycle;
(c) it has clique number 3;
(d) it has chromatic number 4.

Hint for (d): Each vertex has one of the forms (E,E,E), (E,O,O), (O,E,O), or (O,O,E), where E and O mean 'even' and 'odd', respectively. Show that assigning a colour to each form gives a proper colouring.

2.11 Let $G = (V,E)$ be a graph. Let $\mathscr{I}$ be the collection of acyclic subsets of E, that is, subsets A such that the spanning subgraph (V,A) of G contains no cycles. Prove that

(a) $\mathscr{I}$ is non-empty;
(b) $\mathscr{I}$ is downward closed (that is, if $A \in \mathscr{I}$ and $B \subseteq A$, then $B \in \mathscr{I}$;
(c) $\mathscr{I}$ has the *exchange property*, that is, if $A, B \in \mathscr{I}$ and $|A| < |B|$, then there is an element $e \in B \setminus A$ such that $A \cup \{b\} \in \mathscr{I}$.

Deduce that all the maximal elements in $\mathscr{I}$ have the same cardinality, and show that this cardinality is equal to the number of vertices of G minus the number of connected components.

Remark The three conditions in this exercise define the concept of a *matroid*, in terms of its independent sets. Matroids were defined by Hassler Whitney and Takeo Nakasawa (independently) in 1935, to formalise the notion of linear independence in a vector space; but they have many other applications.

As a further exercise, let E be a subset of a vector space, and let $\mathscr{I}$ be the collection of linearly independent subsets of E. Prove that conditions (a) to (c) above hold.

We will not discuss matroids further.

2.12 (a) Find the eigenvalues and describe the eigenvectors of the complete graph K_n.

(b) Let G be the complete bipartite graph $K_{m,n}$. Show that the eigenvalues of G are $\pm \sqrt{mn}$ (each with multiplicity 1) and 0 (with multiplicity $m + n - 2$).

2.13 Let G be a bipartite graph. Show that if λ is an eigenvalue of G, then so is $-\lambda$, having the same multiplicity as λ.

3

Strongly Regular Graphs

In this chapter we consider *strongly regular graphs*, an important class of graphs for many applications, which include both the Shrikhande graph and the graphs associated with Latin squares, all of which play a part in our story.

3.1 The Friendship Theorem

This theorem was proved by Erdős, Rényi and Sós in the 1960s. It was an early success for the methods of algebraic graph theory.

We assume that friendship is an irreflexive and symmetric relation on a set of individuals: that is, nobody is his or her own friend, and if A is B's friend then B is A's friend. (It may be doubtful if these assumptions are valid in the age of social media – but we are doing mathematics, not sociology.) In other words, the situation is described by a graph, in which the vertices are the individuals and two vertices are joined if they are friends.

Theorem 3.1 *In a finite society with the property that any two individuals have a unique common friend, there must be somebody who is everybody else's friend.*

In other words, the configuration of friendships (where each individual is represented by a dot and friends are joined by a line) is as shown in Fig. 3.1.

The proof of Friendship Theorem 3.1 falls into two parts. First, we show that any counterexample to the theorem must be a regular graph. Then, using methods from linear algebra, we show that the only regular graph satisfying the hypothesis is a triangle. Since the triangle satisfies the conclusion of the theorem (each of its vertices is joined to the other two), the theorem is proved.

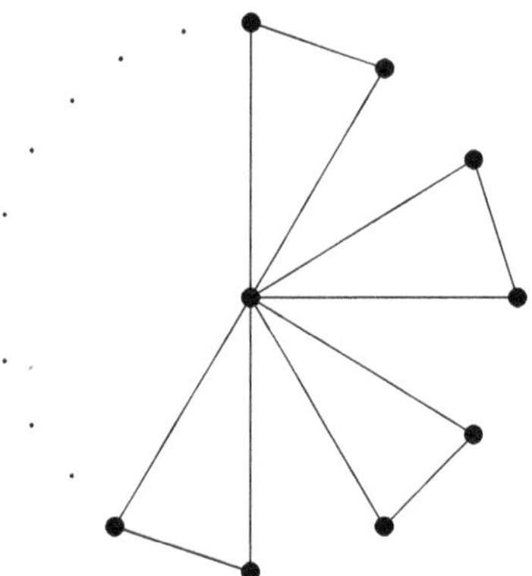

Figure 3.1 The Friendship Theorem

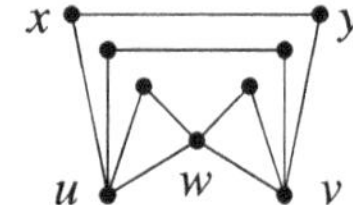

Figure 3.2 A step in the proof

Step 1: A counterexample is regular.

Let G be a graph satisfying the hypothesis of the Friendship Theorem: that is, any two vertices have a unique common neighbour.

First we observe: *Any two non-adjacent vertices of G have the same valency.*

For example, let u and v be non-adjacent. They have one common neighbour, say w. Then u and w have a common neighbour, as do v and w. If x is any other neighbour of u, then x and v have a unique common neighbour y; then obviously y is a neighbour of v, and y and u have a unique common neighbour x. (The sequence (u,x,y,v) is a path.) So we have a bijection from the neighbourhood of u to the neighbourhood of y (Fig. 3.2).

Now we have to show:

If G is not regular, then it satisfies the conclusion of the theorem.

Suppose that a and b are vertices of G with a different valencies. Then a and b are adjacent. Also, any further vertex x has a valency different from that of either a or b, and so is joined to either a or b; but only one vertex, say c, is joined to both.

Suppose that the valency of c is different from that of a. Then any vertex which is not joined to a must be joined to both b and c. But there can be no such vertex, since a is the only common neighbour of b and c. So every vertex other than that of a is joined to a, and the assertion is proved. (The structure of the graph in this case must be a collection of triangles with one vertex identified, as in Fig. 3.1.)

Step 2: A regular graph satisfying the hypotheses must be a triangle.

Let G be a regular graph satisfying the statement of the theorem; let its valency be k.

Let A be the adjacency matrix of G, and let J denote the matrix (with rows and columns indexed by vertices of G) with every entry 1. We have $AJ = kJ$, since each entry in AJ is obtained by summing the entries in a row of A, and this sum is k, by regularity. We claim that

$$A^2 = kI + (J - I) = (k - 1)I + J$$

for the (u, w) entry in A^2 is the sum, over all vertices v, of $A_{uv}A_{vw}$; this term is 1 if and only if $u \sim v$ and $v \sim w$. If $u = w$, then there are k such v (all the neighbours of u), but if $u \neq w$, then by assumption there is a unique such vertex. So A^2 has diagonal entries k and off-diagonal entries 1, whence the displayed equation holds.

Now A is a real symmetric matrix, and so it is diagonalisable. The equation $AJ = kJ$ says that the all-1 vector j is an eigenvector of A with eigenvalue k. Any other eigenvector x is orthogonal to j, and so $Jx = 0$; thus, if the eigenvalue is θ, we have

$$\theta^2 x = A^2 x = ((k - 1)I + J)x = (k - 1)x,$$

so $\theta = \pm \sqrt{k - 1}$.

Thus A has eigenvalues k (with multiplicity 1) and $\pm \sqrt{k - 1}$ (with multiplicities f and g, say, where $f + g = n - 1$).

The sum of the eigenvalues of A is equal to its trace (the sum of its diagonal entries), which is zero since all diagonal entries are zero. Thus $k + f \sqrt{k - 1} + g(- \sqrt{k - 1}) = 0$. So $g - f = k/ \sqrt{k - 1}$.

Now $g - f$ must be an integer; so we conclude that $k - 1$ is a square, say $k = s^2 + 1$, and that s divides k. This is only possible if $s = 1$. So $k = 2$, and the graph is a triangle, as claimed.

3.2 Strongly Regular Graphs

In Step 2 of this proof we considered a special case of the following situation: G is a regular graph with valency k, and the number of common neighbours of two vertices is constant. We generalize this just a little further to obtain the following definition.

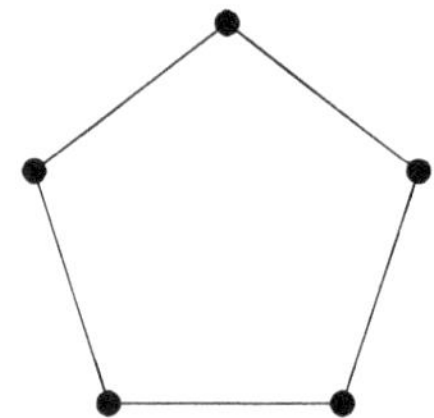

Figure 3.3 The pentagon or 5-cycle

Definition A graph G on n vertices is *strongly regular* if

- G is regular, with valency k (say);
- the number of common neighbours of two vertices v and w is λ if v and w are adjacent and μ if v and w are non-adjacent for two integers λ and μ.

The numbers n, k, λ, μ are called the *parameters* of the strongly regular graph.

The complete and null graphs are strongly regular in a rather trivial way, since there are (respectively) no pairs of non-adjacent vertices, or no pairs of adjacent vertices; we usually disregard these. With this convention, the second part of the proof of the Friendship Theorem asserts that there is no strongly regular graph with parameters $(n, k, 1, 1)$. Slightly less trivially, the disjoint union of t complete graphs, each on $k + 1$ vertices, is strongly regular, with parameters $(t(k + 1), k, k - 1, 0)$; indeed any disconnected strongly regular graph has this form.

A slightly less trivial example is the pentagon or 5-cycle, a regular graph on 5 vertices with valency 2 (Fig. 3.3). Two adjacent vertices have no common neighbour, since the graph contains no triangles; and two non-adjacent vertices have exactly one common neighbour. So the pentagon is strongly regular with parameters $(5, 2, 0, 1)$.

The next result gives two useful properties of strongly regular graphs.

Proposition 3.2 *Let G be a strongly regular graph, with parameters (n, k, λ, μ).*

(a) $k(k - \lambda - 1) = (n - k - 1)\mu$;
(b) the complement $\bar{G}$ of G is also strongly regular, with parameters
     $(n, n - k - 1, n - 2k + \mu - 2, n - 2k + \lambda)$.

Proof For the first part, choose a vertex v and count edges $\{x, y\}$, where x is a neighbour of v and y a non-neighbour. There are k choices for x, each of which is k neighbours, one of which is v and λ are joined to v, so there are $k - \lambda - 1$ further neighbours of x, any of which is a suitable y. But on the other hand, there are $n - k - 1$ non-neighbours of v, each of which is joined to μ neighbours of v by definition.

For the second part, note that each vertex is joined to k others and so not joined to $n - k - 1$. Suppose that v and w are adjacent in $\bar{G}$, so they are not joined in G. There are μ vertices joined to both (in G), and so $k - \mu$ joined to v but not w, and $k - \mu$ joined to w but not v. So there remain $n - 2 - \mu - 2(k - \mu)$ vertices joined to neither (and hence joined to both in $\bar{G}$). The argument in the other case is similar, and we invite you to try it for yourself.

We now describe some further beautiful and important strongly regular graphs.

3.2.1 The Petersen graph

The Petersen graph is one of the most famous objects in graph theory (Fig. 3.4). It has been said that if you have made a new conjecture in graph theory you should test it first on the Petersen graph.

It was first constructed by the Danish mathematician Julius Petersen (1839–1910). He worked in many areas of mathematics including Galois theory and invariant theory but was a pioneer in graph theory. In 1891 he published an important paper proving that a bridgeless cubic graph must contain a 1-factor, a set of pairwise disjoint edges covering every vertex. (In Fig. 3.4, the five edges connecting the outer pentagon to the inner pentagram form a 1-factor.) The Scottish mathematician P. G. Tait showed that the notorious Four Colour Problem is equivalent to the statement that any planar bridgeless cubic graph has a 1-factorization (a partition of the edges into three 1-factors). Petersen's graph (which he published in 1898) showed that not every bridgeless cubic graph has a 1-factorization, thereby showing that Tait's proposed proof of the Four-Colour Theorem would not work.

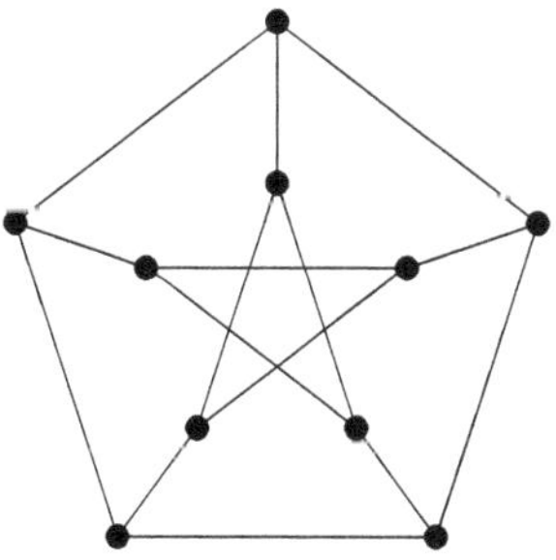

Figure 3.4 The Petersen graph

The Petersen graph can be most briefly described as the complement of the line graph of the complete graph K_5. In other words, the vertices can be labelled with the 2-element subsets of $\{1,2,3,4,5\}$ in such a way that two vertices are joined by an edge if and only if their labels are disjoint. (Readers should find such a labelling of the graph as shown in Fig 3.4.)

The Petersen graph is strongly regular, with parameters $n = 10$, $k = 3$, $\lambda = 0$, $\mu = 1$. Its eigenvalues are 3, 1 and -2, with multiplicities 1, 5 and 4 respectively. The values of λ and μ show that the graph contains no triangles or quadrangles; its girth is 5. A connected graph with diameter 2 and girth 5 is called a *Moore graph of diameter* 2. An early application of graph spectra was the theorem of Hoffman and Singleton.

Theorem 3.3 *The only Moore graphs of diameter 2 are*

- *the 5-cycle;*
- *the Petersen graph;*
- *a unique graph of valency 7 on 50 vertices (now known as the* Hoffman–Singleton graph*);*
- *possibly a graph of valency 57 on 3250 vertices, which has so far defied all attempts at either construction or nonexistence proof.*

For more detail about the Petersen graph, we refer to the book-length treatment of it by Holton and Sheehan [59] (which was in part the inspiration for the present book).

3.2.2 The Clebsch graph

There is in the literature some disagreement about which of two complementary graphs the term *Clebsch graph* applies to. They have 16 vertices; one (which we will call the "Clebsch graph") has 40 edges and valency 5, while its complement has 80 vertices and valency 10. (It is the latter which was originally named the Clebsch graph by Seidel [91].)

The Clebsch graph is a strongly regular graph with parameters $n = 16$, $k = 5$, $\lambda = 0$ and $\mu = 2$. Its eigenvalues are 5, 1 and -3, with multiplicities 1, 10 and 5 respectively.

Alfred Clebsch (1833–72) was born in Königsberg. Despite the fame of this town among graph theorists (the *Königsberg bridge problem* was solved by Euler in what is sometimes regarded as the beginning of graph theory), there is no evidence that Clebsch wrote on graph theory; he was an algebraic geometer and invariant theorist. Among his discoveries was a quartic surface (the *Clebsch quartic*), which contains exactly 16 lines; taking these lines as the vertices of a graph and joining two lines if they intersect gives a 16-vertex regular graph with degree 10, the complement of what we have called the *Clebsch graph*.

The simplest construction of the Clebsch graph is as follows. Start with the 4-dimensional cube Q_4 defined in Section 2.2. Its vertices are all the 4-tuples of zeros and ones, 16 in number; two vertices are joined if they agree in three of their four coordinates and differ in the fourth. The main diagonals of the cube join points which differ in all coordinates. If we add the main diagonals as edges, we obtain the Clebsch graph.

A related construction takes as vertices all the 5-tuples of zeros and ones having at most two ones. Join the all-0 vertex to the five vertices with a single 1; join a vertex with a single 1 to a vertex with two 1s, in this position and one other; and join two vertices with two 1s if and only if their 1s occur in disjoint positions. This construction makes it clear that the set of non-neighbours of the all-0 vertex carries an induced subgraph which is the Petersen graph.

A remarkable property of the Clebsch graph is that the 120 edges of K_{16} can be partitioned into three sets of 40, each of which is a copy of the Clebsch graph. We will discuss this, and the relevance of this for Ramsey's theorem, later in this chapter.

3.2.3 The Schläfli graph

Ludwig Schläfli (1814–95) was a Swiss mathematician who worked on polytopes and algebraic geometry.

The *Schläfli graph* is one of the gems of nineteenth-century algebraic geometry. A general cubic surface, over an algebraically closed field, contains 27 lines, which intersect in 45 points. The graph in question has the 27 lines as its vertices; two vertices are joined by an edge if and only if the lines intersect in a point. (This configuration was discovered by Arthur Cayley, but Schläfli made a deep study of it, and it is now associated with his name. He received the Steiner prize from the Berlin academy for this work.)

Schläfli gave a purely combinatorial construction of this graph; this is referred to as *Schläfli's double-six construction*.

The vertex set of the graph is

$$\{a_i, b_i : 1 \le i \le 6\} \cup \{c_{ij} : 1 \le i < j \le 6\}.$$

If I write c_{ij} with $i > j$, it is to be identified with c_{ji}: in other words, the index of c is an unordered pair.

Edges are as follows:

$$a_i \sim b_j \quad \text{for} \quad i \neq j;$$
$$a_i, a_j, b_i, b_j \sim c_{ij};$$
$$c_{ij} \sim c_{kl} \quad \text{if} \quad \{i,j\} \cap \{k,l\} = \emptyset.$$

Now it is possible to show (and we recommend doing this as an exercise) that the resulting graph is strongly regular, with parameters $n = 27$, $k = 10$, $\lambda = 1$, $\mu = 5$. Its eigenvalues are 10, 1, -5, with multiplicities 1, 20, 6 respectively.

This graph appears in a remarkable theorem of Seidel. We say that a graph has the *strong triangle property* if every edge $\{x,y\}$ is contained in a triangle $\{x,y,z\}$ with the property that any further vertex is joined to exactly one of x, y and z.

Theorem 3.4 *The finite graphs having the strong triangle property are the following:*

- *null graphs (graphs with no edges);*
- *Friendship graphs (those in the conclusion of the Friendship Theorem);*
- *the 3×3 grid $L_2(3)$, or the line graph of $K_{3,3}$;*
- *the complement of the triangular graph $T(6)$;*
- *the Schläfli graph.*

We will need to use this theorem later; we re-state it and give a proof as Theorem 8.2.

3.2.4 The graph

The Shrikhande graph is the topic of the book, so we say no more about it here, except to mention *Shrikhande's Theorem*, which we will prove in the next section. This states that a strongly regular graph with parameters $(n^2, 2(n-1), n-2, 2)$ is isomorphic either to $L_2(n)$ (defined in the following) or to the Graph (with $n = 4$).

3.2.5 The three Chang graphs

The situation for the triangular graph $T(n) = L(K_n)$ is similar to that for the square lattice graph, but a bit more complicated. The first part of the next theorem was proved by Hoffman; the exceptions were worked out by Chang.

Theorem 3.5 *Let G be a graph cospectral with $T(n) = L(K_n)$, or equivalently, let G be a strongly regular graph with the same parameters as $T(n)$.*

- *If $n \neq 8$, then G is isomorphic to $T(n)$.*
- *For $n = 8$, there are just three exceptional graphs (up to isomorphism).*

The simplest description of the three Chang graphs was found by Seidel and involves the operation of *Seidel switching*, which we will describe in more detail in Section 10.2. Let X be a subset of the vertex set V of a graph G. The operation σ_A of switching G with respect to X consists of replacing all edges between X and its complement $V \setminus X$ by non-edges, and all non-edges by edges, while leaving edges and non-edges within X or within $V \setminus X$ unchanged.

The vertex set of $T(8)$ can be regarded as the set of 28 edges of the complete graph K_8. So any subset of the vertex set can be regarded as the edge set of a graph on 8 vertices. Seidel showed that the three Chang graphs are obtained by switching $T(8)$ with respect to the following sets:

- a 1-factor;
- an 8-cycle;
- the disjoint union of a 3-cycle and a 5-cycle.

We have been unable to find any biographical information about Chang Li-Chien, who found and characterized these graphs in 1959.

3.2.6 Strongly edge triangle regular graphs

There is another notion called *strongly edge triangle regular graphs*, which is defined as follows.

Definition A graph G is *edge triangle regular* if all of its edges have the same triangle number, where *triangle number* of an edge is the number of triangles in which it is present. An edge triangle regular graph, which is also regular, is called strongly edge triangle regular.

The following theorem from [83] establishes the relation between strongly regular graphs and strongly edge triangle regular graphs.

Theorem 3.6 *A graph G is strongly regular if and only if both G and $\bar{G}$ are strongly edge triangle regular.*

Proof Let G be a strongly regular graph with parameters n, k, λ and μ. Then $\bar{G}$ is strongly regular with parameters $n, n - k - 1, n - 2k + \mu - 2$ and $n - 2k + \lambda$. Hence, in G, $d(u) = k$ and $t(e) = \lambda$ for every vertex u and edge e, and in $\bar{G}$, $d(u) = n - k - 1$ and $t(e) = n - 2k + \mu - 2$ for every vertex u and edge e. Thus, G and $\bar{G}$ are strongly edge triangle regular. Conversely, let G and $\bar{G}$ be strongly edge triangle regular graphs and let $d(u) = r$ for every vertex u in G, $t(e) = t$

for every edge e in G and $t(e) = \bar{t}$ for every edge e in G. Then in G, any two adjacent vertices have t common neighbours and any two non-adjacent vertices have $2r + \bar{t} - n + 2$ common neighbours. Therefore, G is strongly regular with parameters n, r, t and $2r + \bar{t} - n + 2$, hence, the theorem.

3.3 Square Lattice and Triangular Graphs

Two famous and important infinite families of strongly regular graphs are the following.

(a) Let m be a positive integer. The *square lattice graph* $L_2(m)$ has m^2 vertices, which are labelled with the ordered pairs of integers from the set $\{1, \ldots, m\}$. Two distinct vertices (x, y) and (u, v) are adjacent if and only if $x = u$ or $y = v$ (but not both). This graph can be represented by the points in an $m \times m$ square grid, with two elements joined if and only if they are in the same row or column of the grid (Fig. 3.5(a)). It is strongly regular with parameters $(m^2, 2(m-1), m-2, 2)$.

Alternatively, it is the line graph of the complete bipartite graph $K_{m,m}$: if the vertices of the complete bipartite graph are labelled $\{a_1, \ldots, a_m\}$ in one bipartite block and $\{b_1, \ldots, b_m\}$ in the other, then the edge $\{a_x, b_y\}$ corresponds to the vertex (x, y) of $L_2(4)$.

(b) Let m be a positive integer. The vertex set of the triangular graph $T(m)$ is the set of all 2-element subsets of $\{1, \ldots, m\}$ with two vertices adjacent if the corresponding sets intersect in one point (see Fig. 3.5(b)). Alternatively, it is the line graph of the complete graph K_m, the 2-set $\{x, y\}$ corresponding to the edge joining x to y. It is strongly regular, and its parameters are $(m(m-1)/2, 2(m-2), m-2, 4)$.

These graphs are nearly always determined by their parameters, in the sense that any strongly regular graph with the same parameters is isomorphic to one

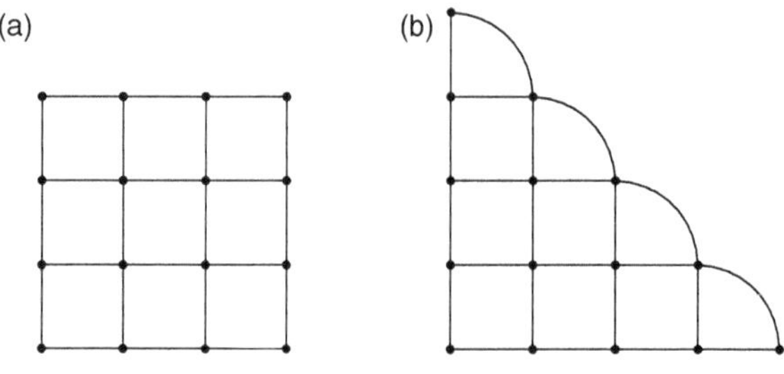

Figure 3.5 (a) The graph $L_2(4)$ (b) The graph $T(6)$

of these, except for a very few cases. This is very far from being true for strongly regular graphs with arbitrary parameters. The pioneering result showing this, which was the foundation for much that followed, is due to Shrikhande, Hoffman and Chang:

Theorem 3.7 *(a) If $m \neq 4$, then a strongly regular graph having the same parameters as $L_2(m)$ is isomorphic to $L_2(m)$. For $m = 4$, there is just one exceptional graph with the same parameters as $L_2(4)$, namely the* Shrikhande graph.

(b) If $m \neq 8$, then a strongly regular graph having the same parameters as $T(m)$ is isomorphic to $T(m)$. For $m = 8$, there are three exceptional graphs with the same parameters as $T(8)$, the three Chang graphs.

3.4 Shrikhande's Theorem

In this section we give a proof of Shrikhande's Theorem, using the sort of techniques Shrikhande pioneered. Later we will see other approaches to the theorem.

Recall the graph $L_2(n)$, whose vertices form an $n \times n$ square array, two vertices being joined if and only if they lie in the same row or the same column of the array. This graph is strongly regular, with parameters $(n^2, 2(n-1), n-2, 2)$.

Theorem 3.8 *Let G be a strongly regular graph. Suppose that its parameters are $(n^2, 2(n-1), n-2, 2)$ for some integer $n > 4$. Then G is isomorphic to $L_2(n)$.*

Proof Choose a vertex v. Let N be the set of neighbours of v, and H the induced subgraph on the set N. The graph H has $2(n-1)$ vertices and valency $n-2$.

Choose two nonadjacent vertices $x, y \in N$. Then x and y have two neighbours in G (since $\mu = 2$), one of which is v; so they have at most one common neighbour in N.

Suppose that there is a common neighbour. Then, in N, there is one vertex joined to x and y, $n-3$ vertices joined to x but not y, $n-3$ vertices joined to y but not x and hence $2(n-1) - 2 - 2(n-3) - 1 = 1$ vertex joined to neither x nor y; let z be this vertex. The argument reverses: if two vertices in N are not adjacent, and some vertex is joined to neither, then there is a vertex joined to both.

Suppose that in N, we have three pairwise non-adjacent vertices x, y, z. There are two cases:

- There might be no vertex joined to all three. Then three vertices joined to two of x, y, z, and $3(n-4)$ vertices joined to just one of them. Since $|N| = 2(n-1)$, we have

$$3 + 3 + 3(n-4) \leq 2(n-1),$$

giving $n \leq 4$, contrary to hypothesis.

- There may be a vertex w joined to all three. Then no vertex is joined to two of x, y, z, and there are $3(n-3)$ vertices joined to just one of them. So

$$1 + 3 + 3(n-3) \leq 2(n-1),$$

so that $n \leq 3$, again contrary to hypothesis.

So two non-adjacent vertices in N have no common neighbours. It follows that H consists of two complete graphs of size $n - 1$. Since the vertex v was arbitrary, we see that each vertex lies in two "grand cliques" of size n. Counting incidences between vertices and grand cliques, we see that the number of grand cliques is $n^2 \cdot 2/n = 2n$.

Regarding the grand cliques as lines of a geometry whose points are the vertices, we see that the geometry contains no triangles. For if three grand cliques C_1, C_2, C_3 satisfy $C_1 \cap C_2 = \{v\}$, then $v \notin C_3$, and so the neighbourhood of v contains an edge (in C_3) joining points in the cliques $C_1 \setminus \{v\}$ and $C_2 \setminus \{v\}$. But this contradicts the structure of the neighbourhood of v.

Thus, if C is a grand clique, then the second grand cliques through the vertices in C are pairwise disjoint, and cover the whole vertex set without overlaps. Repeating the argument with one of these grand cliques replacing C, we see that the grand cliques fall into two families each containing n cliques forming a partition of the vertex set. Taking these as the rows and columns of a square array, we have identified G with $L_2(n)$.

It is not difficult to see that if $n = 3$, a strongly regular graph with the parameters of $L_2(3)$ is isomorphic to $L_2(3)$: we leave this as an exercise for the reader. Shrikhande also analysed the remaining case $n = 4$:

Theorem 3.9 *There are just two graphs up to isomorphism which are strongly regular with parameters* $(16, 6, 2, 2)$.

One of these graphs is $L_2(4)$; the other is the graph now known as the *Shrikhande graph*.

Proof We begin in a similar way to the previous proof. Let N be the set of neighbours of a vertex v, and H the induced subgraph on N. Then H is a regular graph on 6 vertices with valency 2; so H is either the disjoint union of two triangles or a single 6-cycle. Call a vertex of Type A if its neighbourhood is a disjoint union of two triangles, and of Type B if it is a 6-cycle.

If v has Type A, then for every neighbour w of v, the two common neighbours of v and w are joined by an edge, since these are the neighbours of w in the graph H. On the other hand, if v has Type B, then for every neighbour w of v, the two common neighbours of v and w are not joined. It follows that if v has Type A, then every neighbour of v has Type A. Since the graph is connected, we see that in this case every vertex has Type A. We conclude that either all vertices have Type A or all vertices have Type B.

In the first case, just as in the proof of the preceding theorem, there are $2 \cdot 4 = 8$ grand cliques of size 4, and as in that proof, the graph is isomorphic to $L_2(4)$.

So suppose that the second case holds, so that every vertex has Type B. We note that in this case the neighbourhood of a vertex contains no 3-clique, so the graph G contains no 4-clique.

Choose a vertex v, and let (a,b,c,d,e,f) be the 6-cycle in the neighbourhood N of v (Fig. 3.6).

Let P be the set of non-neighbours of v. We have to identify the nine vertices in P and the edges between N and P as well as those within P.

Each vertex in P has two neighbours in N, since $\mu = 2$. Two vertices of N at distance 2 in the 6-cycle on N (say a and c) already have two common neighbours (v and b). So the possible subsets of N which can be neighbours of

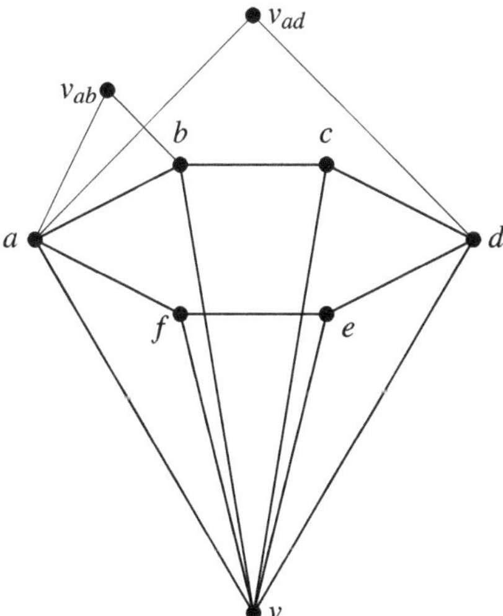

Figure 3.6 Part of the Shrikhande graph

vertices in P are the six edges and the three antipodal pairs in the 6-cycle. Since there are nine vertices in P, we see that each is uniquely identified by its two neighbours in N. Label the vertices as shown in Fig. 3.6, so that for example $v_{ab} \in P$ is joined to a and b in N.

Now consider the neighbours of a. These are v, b, f, v_{ab}, v_{af} and v_{ad}. We already have four edges within this set. So we must have edges between v_{ad} and v_{ab} and between v_{ad} and v_{af}; but v_{ab} and v_{af} are not joined.

The induced subgraph on P has valency 4. We have now identified four neighbours for the vertices corresponding to antipodal pairs (for example, v_{ad} is joined to v_{ab}, v_{af}, v_{cd} and v_{de}); so these vertices lie on no further edges. The vertex v_{ab} is joined to v_{ad} and v_{be} but not to v_{af} or v_{bc}. So its remaining neighbours must be two of v_{cd}, v_{de} and v_{ef}. If it is joined to v_{de}, then $\{v_{ab}, v_{de}, v_{ad}, v_{be}\}$ would be a 4-clique, contrary to our earlier observation. Thus it is joined to v_{ad}, v_{be}, v_{cd} and v_{ef}. So we have identified the neighbours of v_{ab} (and similarly the remaining vertices of P), and the proof is complete.

Let us now reverse the procedure to give a construction of the graph, rather different from the one we saw earlier.

Construction Let H be a hexagon (a 6-cycle graph). Let V, E, D be the sets of vertices, edges and long diagonals of H respectively and let $*$ be an additional vertex. Build a graph as follows:

- The vertex set is $\{*\} \cup V \cup E \cup D$.
- $*$ is joined to every vertex in V.
- A vertex in V is joined to its neighbours in the hexagon and to the elements of E and D which contain it.
- An element of E is joined to the two elements of D which meet it and the two elements of E which are two steps from it in the hexagon.

We leave to the reader the job of showing that this graph is strongly regular with parameters $(16, 6, 2, 2)$.

We note several unexpected bonuses from this construction. To check that a graph is isomorphic to the Shrikhande graph, we can see whether it can be built step by step from a vertex, as in this construction. Also, we will see later that it gives us another way to study the automorphisms of the Shrikhande graph.

3.5 Partitions into Given Graphs

The complete graph K_{10} has 45 edges and degree 9, while the Petersen graph has 15 edges and degree 3. So it seems possible that the edge set of K_{10} could be partitioned into three copies of the Petersen graph. The following theorem, due to Allen Schwenk, shows that such a partition is not possible. It is a nice application of algebraic graph theory.

Theorem 3.10 *The edge set of K_{10} cannot be partitioned into three copies of the Petersen graph. Indeed, if two disjoint copies of the edge set of the Petersen graph are removed, the graph remaining is bipartite.*

Proof We have seen that the adjacency matrix of the Petersen graph has eigenvalues 5, 1 and -2, with eigenvalues 1, 5 and 4 respectively. The all-1 vector u is an eigenvector with eigenvalue 3, and the remaining eigenvectors are orthogonal to u.

Suppose that $J - I = A(K_{10}) = A_1 + A_2 + A_3$, where J is the all-1 matrix and, for $i = 1, 2, 3$, $A_i = A(G_i)$, where G_1 and G_2 are edge-disjoint copies of the Petersen graph and G_3 consists of the remaining edges.

Thus $\mathbb{R}^{10}$ has a 5-dimensional subspace W_1 consisting of eigenvectors of $A(G_1)$ with eigenvalue 1, which is contained in the 9-dimensional space orthogonal to the all-1 vector u, similarly for $A(G_2)$, with eigenspace W_2.

We have $\dim(W_1 + W_2) \le 9$; so $\dim(W_1 \cap W_2) \ge 1$. Let w be a non-zero vector in $W_1 \cap W_2$. Then

$$0 = Jw = Iw + A_1w + A_2w + A_3w = 3w + A_3w;$$

so $A_3w = -3w$, and we conclude that G_3 is not isomorphic to the Petersen graph (which does not have an eigenvalue -3).

In fact, by Theorem 2.8, a regular graph with degree 3 having an eigenvalue -3 is bipartite (the positions of the positive and negative entries in the eigenvector give a bipartition). So our argument shows that if G_1 and G_2 are edge-disjoint copies of the Petersen graph in K_{10}, then the complement of their union is bipartite. $\square$

Now consider the Clebsch graph, which has 40 edges and degree 5. Since K_{16} has 120 edges and degree 15, we could ask whether the edge set of K_{16} can be partitioned into three copies of the Clebsch graph. This time, the answer is different; the partition does exist, and this fact is important in Ramsey theory. Before proving this, we recall that the Clebsch graph is strongly regular with parameters $(16, 5, 0, 2)$; it has eigenvalues 5 (multiplicity 1) 1 (multiplicity 10) and -3 (multiplicity 5).

Proposition 3.11 *A strongly regular graph with eigenvalues $(16, 5, 0, 2)$ is isomorphic to the Clebsch graph.*

Proof Select a vertex v, and let X and Y be the sets of neighbours and non-neighbours of v. Then $|X| = 5$, and X contains no edges (since $\lambda = 0$); and $|Y| = 10$. Each vertex in Y has two neighbours in X, since $\mu = 2$; and any two vertices in X have a common neighbour v, and so one further common neighbour in Y. Thus the sets $N(y) \cap X$ for $y \in Y$ are all the 2-element subsets of X, which we can use to label them: $y = y_{ab}$ if $N(y) \cap X = \{a,b\}$.

Vertices y_{ab} and y_{ac} have a common neighbour a so are not joined. Thus y_{ab} is joined only to vertices whose labels are disjoint from $\{a,b\}$. Since y_{ab} has three further neighbours, and there are three 2-sets disjoint from $\{a,b\}$, we have found all edges, and proved the uniqueness. Note that the induced subgraph on Y is isomorphic to the Petersen graph. $\qquad\square$

Theorem 3.12 *The edge set of K_{16} can be partitioned into three copies of the Clebsch graph. In fact, if two disjoint copies of the edge set of the Clebsch graph are removed from K_{16}, the resulting graph is isomorphic to the Clebsch graph.*

Proof We will give a construction later, in Exercise 6.4(d); in the meantime, readers might wish to try to find the partition for themselves.

As in the preceding theorem, suppose that $J - I = A(K_{16}) = A_1 + A_2 + A_3$, where for $i = 1,2,3$, $A_i = A(G_i)$, where G_1 and G_2 are edge-disjoint copies of the Clebsch graph and G_3 consists of the remaining edges. So each of A_1 and A_2 has a 10-dimensional space of eigenvectors with eigenvalue 1, contained in the 15-dimensional space orthogonal to the all-1 vector u. These two spaces have intersection of dimension (at least) 5. If v is a vector in this space, then $A_1v = A_2v = Iv = v$, $Jv = 0$; so $A_3v = -3v$. Thus A_3 has eigenvalue -3 (with multiplicity 5) in addition to 5 (with multiplicity 1).

Let $r_1,\ldots,r_{10}$ be the remaining eigenvalues of A_3. Since the trace of A_3 is 0, we have $5 + 5 \cdot (-3) + r_1 + \cdots + r_{10} = 0$, so

$$r_1 + \cdots + r_{10} = 10.$$

Moreover, the trace of A_3^2 is 80, since all of its diagonal elements are 5; thus $25 + 5 \cdot 9 + r_1^2 + \cdots + r_{10}^2 = 80$, so

$$r_1^2 + \cdots + r_{10}^2 = 10.$$

From this we see that

$$(r_1 - 1)^2 + \cdots + (r_{10} - 1)^2 = 0,$$

so $r_1 = \cdots = r_{10} = 1$. So A_3 has the same spectrum as the Clebsch graph: eigenvalues 5, 1, -3 with multiplicities 1, 10 and 5 respectively.

But the eigenvalues determine the strongly regular graph parameters, and hence G_3 is isomorphic to the Clebsch graph (by our uniqueness proof). $\qquad\square$

This decomposition was found by Greenwood and Gleason [52], in connection with Ramsey's theorem, which we discuss in the next section.

Is there anything similar for the Shrikhande graph? This graph has 48 edges and degree 6, whereas the complete graph K_{16} has 120 edges and degree 15; so we cannot partition K_{16} into copies of the Shrikhande graph. However, suppose that we double every edge of K_{16} to obtain the multigraph $K_{16}^{\{2\}}$, with 240 edges and degree 30. We can then ask whether $K_{16}^{\{2\}}$ can be partitioned into five copies of the Shrikhande graph.

We actually consider a more general problem. Since the graph $L_2(4)$ has the same number of edges and the same degree as the Shrikhande graph, we can set six different problems:

Problem For $0 \leq i \leq 5$, is it possible to partition the edges of $K_{16}^{\{2\}}$ into i copies of the Shrikhande graph and $5 - i$ copies of $L_2(4)$?

We will settle three of these six problems later, in Section 6.8. The remaining three are left to the ingenuity of our readers.

For further extensions of these problems we refer to [32, 102].

3.6 Ramsey's Theorem

Suppose that we colour the edges of K_{16} with three colours, say red, green and blue, so that the edges in the first copy of the Clebsch graph are red, those in the second are green and those in the third are blue. Then there are no monochromatic triangles (triangles with all edges of the same colour).

This is relevant to an important theorem proved by Frank Ramsey in 1930 [84]. We generalize this example in three ways:

(a) We allow any number r of colours, not just three.
(b) We are interested in the existence of a monochromatic set of size l, not necessarily 3, that is a set such that all edges within it have the same colour.
(c) Finally, instead of taking the complete graph and colouring edges, we take the *complete k-uniform hypergraph* (whose *hyperedges* are all the k-element subsets of the vertex set) and colour the hyperedges.

Now Ramsey's theorem is the following:

Theorem 3.13 *Given positive integers k, l, r with $l \geq k$, there is a positive integer N with the property that if the hyperedges of the k-uniform hypergraph on N vertices are coloured with r colours, there is a monochromatic set of size l.*

Note that if an integer N has this property, then so does any larger integer. The *Ramsey number* $R(k,l,r)$ is defined to be the smallest number N with the property of the theorem.

Example $R(2,3,2) = 6$. This means that if the edges of K_6 are coloured red and blue, there is a monochromatic triangle. This is the famous *party problem*: in any group of 6 people at a party, there are either 3 mutual acquaintances, or 3 mutual strangers. Moreover, 5 people are not enough to force this conclusion.

To show that 6 are enough, choose a vertex a. Since a lies on five edges, there must be at least three of them with the same colour; say (without loss of generality) $\{a,b\}$, $\{a,c\}$ and $\{a,d\}$ are red. Now, if any two of b,c,d are joined by a red edge, they form a red triangle with a; if not, then $\{b,c,d\}$ is a blue triangle.

To show that five are not enough, colour the edges of a pentagon red and the diagonals blue. No monochromatic triangle is formed.

Note that to establish the exact value of this Ramsey number, we had to do two things: show that six vertices are enough to force a monochromatic triangle, but five are not enough. Here is another example.

Proposition 3.14 $R(2,3,3) = 17$.

Proof For the first task, colour the edges of K_{17} red, blue and green arbitrarily; we must find a monochromatic triangle. Choose a vertex a; there must be at least six edges of the same colour containing a, so let us suppose that $\{a,x\}$ is green for $x \in X$, where $|X| = 6$. If X contains a green edge, we have a green triangle, and we are done; if not, the edges in X are coloured red and blue, and the previous case $R(2,3,2) = 6$ shows that there is a monochromatic triangle.

For the second task, take K_{16} with the edges partitioned into three copies of the Clebsch graph, and give one of the three colours to each copy. Since the Clebsch graph contains no triangle, we have a colouring without a monochromatic triangle. $\qquad\square$

We refer to the book by Graham, Rothschild and Spencer [51] for further information about Ramsey theory and Ramsey numbers. Up-to-date information on Ramsey numbers (including upper and lower bounds) can be found in [85]. In particular, at the time of writing, the upper and lower bounds for the best possible value of $R(4,2,3)$ are 62 and 51. Calculating the exact values of Ramsey numbers is an extremely difficult problem, and only a small handful are known precisely. The combinatorialist Paul Erdős gave his view of the difficulty of this problem. The value of $R(2,4,2)$ is known to be 18, and $R(2,l,2)$ is unknown for $l > 4$. Erdős said that if aliens invaded Earth and threatened to destroy the planet unless we gave them the value of $R(2,5,2)$, we could put every mathematician and computer to work on the problem and maybe solve it; but if they asked for $R(2,6,2)$, our only chance would be to get them before they got us!

Exercises

3.1 Show that there is a unique strongly regular graph with parameters $(9,4,1,2)$.

3.2 Suppose that the strongly regular graph G is isomorphic to its complement. Show that the parameters (n,k,λ,μ) satisfy $n = 4\mu + 1$, $k = 2\mu$, $\lambda = \mu - 1$.

3.3 Suppose that the strongly regular graph G has irrational eigenvalues. Show that the parameters (n,k,λ,μ) satisfy $n = 4\mu + 1$, $k = 2\mu$, $\lambda = \mu - 1$.

3.4 Show that the graph constructed in Fig. 3.6 is indeed strongly regular, and that it is isomorphic to the Shrikhande graph given by the construction in Section 2.6.

3.5 Show that the graph shown in Fig. 3.7 is a subdivision of K_5 and a subgraph of the Shrikhande graph as defined in Fig. 3.6.

Deduce from Kuratowski's theorem that the Shrikhande graph is not planar.

Does the Shrikhande graph contain a subdivision of $K_{3,3}$?

3.6 Find two edge-disjoint copies of the Petersen graph in K_{10}, and verify that the complement of their union is indeed bipartite.

3.7 Show that there is a function f such that if the edges of K_n are coloured with r colours and $n \geq f(r)$, then there must be a monochromatic triangle.

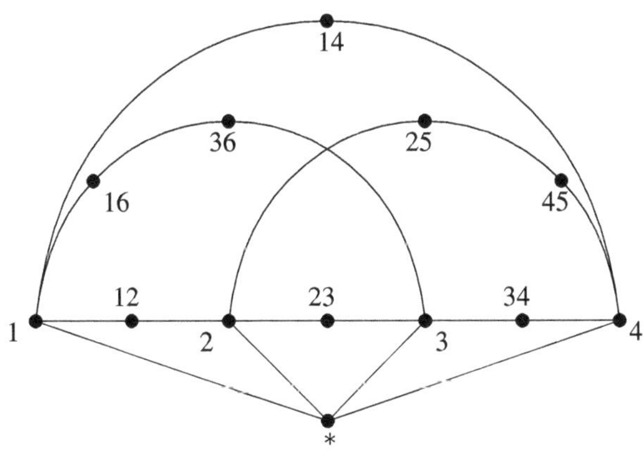

Figure 3.7 A subdivision of K_5 in the Shrikhande graph

PART III

Properties of the Shrikhande Graph

4

Spectrum and Automorphism Group

4.1 Spectrum

We saw earlier that the Shrikhande graph is a strongly regular graph, with parameters $(16, 6, 2, 2)$. In Chapter 3, we saw that this implies that the adjacency matrix A satisfies the two equations

$$AJ = 6J, \qquad A^2 = 4I + 2J,$$

where I and J are the identity and all-1 matrices of order 16.

The first equation shows that the all-1 vector j is an eigenvector of A with eigenvalue 6. From the theory of real symmetric matrices we know that any other eigenvector v is orthogonal to j, and hence if the corresponding eigenvalue is θ, we have

$$\theta^2 v = A^2 v = 4v,$$

so $\theta = \pm 2$.

If the multiplicities of the eigenvalues $+2$ and -2 are f and g respectively, we see that

$$f + g = 15,$$
$$2f - 2g = -6.$$

The first equation holds because the other eigenvalue 6 has multiplicity 1. The second equation uses the fact that the trace of a matrix is the sum of its eigenvalues, and A has trace 0 since all its diagonal elements are 0.

The solution of these two equations is $f = 6$, $g = 9$. So the eigenvalues are 6, 2 and -2 with multiplicities 1, 6, 9, respectively. Thus the claims we made after Hoffman's ratio bound (Theorem 2.10) are verified.

59

4.2 Permutation Group Properties

In this section we examine some properties of a *permutation group* (a group acting on a set), and see how they work out for the Shrikhande graph.

Let Γ be a group whose elements are permutations of a set X. We assume that the action is "on the right", so that the effect of acting on x by γ and then by δ is the same as the effect of acting by $\gamma\delta$ (the composition in the group). We write $x\gamma$ for the image of x under the permutation γ. So by our convention,

$$(x\gamma)\delta = x(\gamma\delta).$$

(If we wrote the image of x under γ in function notation as $\gamma(x)$, we would have the less natural equation $(\gamma\delta)(x) = \delta(\gamma(x))$.)

The theory of permutation groups is the oldest part of algebra, having perhaps begun with Galois in the nineteenth century. We need very little of this.

Definition Let Γ be a group of permutations on a set X. Define an equivalence relation $\sim$ by the rule that $x \sim y$ if there exists $\alpha \in \Gamma$ with $x\alpha = y$. The equivalence classes of this equivalence relation are called the *orbits* of Γ; if there is just a single orbit, we say that Γ acts *transitively* on X (or, more briefly, Γ is *transitive* on X).

The definition requires comment. Why is $\sim$ an equivalence relation?

- It is reflexive, because $xe = x$ for all x, where x is the identity element of Γ.
- It is symmetric, because if $x\alpha = y$, then $y\alpha^{-1} = x$.
- It is transitive, because if $x\alpha = y$ and $y\beta = z$, then
  $x(\alpha\beta) = (x\alpha)\beta = y\beta = z$, by the rule for composition of mappings.

Thus the three parts of the definition of an equivalence relation follow from three group axioms: closure, identity and inverses. (The associative law is not necessary since composition of mappings is always associative.)

This analysis leads to a version of Lagrange's theorem in group theory, called the *Orbit-Stabilizer Theorem*:

Theorem 4.1 *Let the group Γ act on a set X. Let O be an orbit of Γ, x a point in O and Δ the* stabilizer *of x (the set of elements of Γ which fix x). Then Δ is a subgroup of Γ, and*

$$|O| \cdot |\Delta| = |\Gamma|.$$

Proof The verification that Δ is a subgroup is straightforward: if two elements of Γ fix x, then so do their product and inverses. Now we can show that for any $y \in O$, the set of elements of Γ mapping x to y form a right coset of Δ, and the result follows since right cosets of a subgroup contain the same number of elements as the subgroup. □

If Γ is a permutation group on X, then Γ has a natural action on X^2, the set of ordered pairs of elements of Ω, by the rule

$$(x,y)\alpha = (x\alpha, y\alpha).$$

The number of orbits in this action is the *rank* of the permutation group Γ. If $|X| > 1$, then the rank of a permutation group on X is at least 2, since the pairs (x,x) and (x,y) (where $y \neq x$) cannot lie in the same orbit. If the rank is 2, then we can map any ordered pair of distinct elements to any other by a permutation in Γ: in this case, Γ is said to be *2-transitive*.

Now suppose that X is the vertex set of a graph G, and that Γ consists of automorphisms of G (in other words, $\Gamma \leq \mathrm{Aut}(G)$). We call the action *vertex-transitive* or *edge-transitive* if the action on vertices or edges is transitive. In the case where $\Gamma = \mathrm{Aut}(G)$, we say that G is a *vertex-transitive graph* or an *edge-transitive graph*.

This terminology can be extended to other configurations in the graph; for example, we can speak of a *non-edge transitive graph*.

A very strong symmetry condition on the graph is that its automorphism group is transitive on vertices, on ordered edges (that is, ordered pairs (v,w) where $\{v,w\}$ is an edge) and on ordered non-edges. In this case, the automorphism group has just three orbits on ordered pairs of vertices, so has rank 3 per our earlier terminology; we describe Γ as a *rank 3 graph*.

Proposition 4.2 *A rank 3 graph is strongly regular.*

Proof If an automorphism α maps (x,y) to (x',y'), then it maps the common neighbours of x and y to the common neighbours of x' and y', so the numbers of such common neighbours are equal. So, if G is a rank 3 graph, then the number of common neighbours of x and y depends only on whether x and y are equal, adjacent or non-adjacent. □

The converse is false. In fact, we have:

Proposition 4.3 *The Shrikhande graph is the smallest graph which is strongly regular but not rank 3.*

We have seen the proof of Shrikhande's Theorem 3.8 that the neighbourhood of a vertex v is a 6-cycle, and if w is not joined to v, then the common neighbours of v and w form either an edge or a long diagonal of this 6-cycle. So there are two types of non-edges $\{v,w\}$: the two common neighbours of v and w can be either an edge or a non-edge. It is not possible to map a pair of one type to a pair of the other.

Showing that all smaller strongly regular graphs are rank 3 graphs involves a certain amount of labour, which we do not give here.

In fact the automorphism group of the Shrikhande graph is a rank 4 group: it is transitive on the four types of pairs of vertices (equal, adjacent, non-adjacent but common neighbours form an edge, and non-adjacent but common neighbours form a non-edge).

4.3 Cayley Graphs

One important source of vertex-transitive graphs consists of Cayley graphs.

Arthur Cayley showed that any group Γ can be represented as a group of permutations; that is, its elements can be mapped bijectively to permutations of a set X in such a way that the group operation becomes a composition of permutations. If Γ is a finite group of order n, then we can take X to be a set of n elements, so that Γ is embedded as a subgroup of the symmetric group S_n.

Cayley's construction was as follows. The *Cayley table* of a group Γ of order n is the $n \times n$ array with rows and columns indexed by Γ, so that in row α and column β we have the element $\gamma = \alpha\beta$, which is the result of applying the group operation to α and β: $\gamma = \alpha \circ \beta$. Now it follows from the group axioms that each row and each column of this table is a permutation of the column or row labels. We will use columns. So associated with the element β of Γ is the permutation p_β which maps α to $\alpha \circ \beta = \gamma$. The group property is easy to check: applying p_β and then p_γ, the element α is mapped to

$$(\alpha p_\beta)p_\gamma = (\alpha\beta)p_\gamma = (\alpha\beta)\gamma = \alpha(\beta\gamma) = \alpha p_{\beta\gamma},$$

so the result is the permutation corresponding to $\beta\gamma$.

The next step that Cayley took was to construct a graph which admits the group Γ as a group of automorphisms. Let S be a subset of Γ. We make two assumptions; shortly we will see what happens if these assumptions are relaxed.

- The identity element of Γ is not in S;
- S is closed under inversion: that is, if $\alpha \in S$, then $\alpha^{-1} \in S$.

Definition Let Γ be a group and S an inverse-closed subset of Γ not containing the identity. The *Cayley graph* $\mathrm{Cay}(\Gamma, S)$ is the graph with vertex set Γ and an edge $\{\alpha, \beta\}$ whenever $\alpha, \beta \in \Gamma$ and $\beta = \sigma\alpha$ for some $\sigma \in S$.

Note that the two conditions on S guarantee that we have a simple graph: no loops because the identity is not in S and undirected edges because if $\beta = \sigma\alpha$ then $\alpha = \sigma^{-1}\beta$.

Theorem 4.4 *Let Γ be a group and S an inverse-closed subset of Γ not containing the identity.*

(a) $\text{Cay}(\Gamma,S)$ *is connected if and only if S generates Γ (that is, lies in no proper subgroup).*
(b) Γ, acting on itself by right multiplication as in Cayley's theorem, is a group of automorphisms of $\text{Cay}(\Gamma,S)$.
(c) $\text{Cay}(\Gamma,S)$ *is vertex-transitive.*

Proof (a) By a simple induction, any vertex that can be reached by a path from the identity is a product of elements in S, and so lies in the group generated by S.

(b) Let $e = \{\alpha, \sigma\alpha\}$ be an edge of $\text{Cay}(\Gamma,S)$. Now the permutation p_β maps α to $\alpha\beta$ and $\sigma\alpha$ to $(\sigma\alpha)\beta = \sigma(\alpha\beta)$, by the associative law; so the image of e is an edge.

(c) To map vertex α to vertex β, we can apply the automorphism $p_{\alpha^{-1}\beta}$. $\quad\square$

Proposition 4.5 *The Shrikhande graph is a Cayley graph.*

Proof We use the description in Section 2.6. The vertices of Γ form the group

$$\Gamma = \{(x,y,z) \in (\mathbb{Z}_4)^3 : x + y + z = 0\},$$

and the edges are those of the Cayley graph $\text{Cay}(\Gamma,S)$, where

$$S = \{(0,1,-1),(0,-1,1),(-1,0,1),(1,0,-1),(1,-1,0),(-1,1,0)\}. \quad \square$$

We observed that every Cayley graph is vertex-transitive. Cayley graphs comprise a very important subclass of vertex-transitive graphs; but not every vertex-transitive graph is a Cayley graph.

Proposition 4.6 *The Petersen graph is not a Cayley graph.*

Proof If the Petersen graph G is a Cayley graph of Γ, then Γ has order 10, and so contains an element of order 2. Thus, the Petersen graph would have an automorphism of order 2 with no fixed points. But we claim that any automorphism of order 2 has a fixed point. Let α be an automorphism of order 2, and suppose that α interchanges x and y. If x and y are not joined, then their unique common neighbour z is fixed by α. But if x and y are joined, then the neighbours of x are interchanged with the neighbours of y, and these are mutually non-adjacent, so the previous argument applies. $\quad\square$

We have seen that Γ is a subgroup of the automorphism group of any of its Cayley graphs. Usually it is a proper subgroup. If it happens that $\text{Aut}(\text{Cay}(\Gamma,S)) = \Gamma$, then we say that $\text{Cay}(\Gamma,S)$ is a *graphical regular representation* of Γ. The groups which have graphical representations have been classified; many authors contributed to this, the final step being taken by Godsil [48]. They are the abelian groups which are not elementary abelian 2-groups, the generalized dicyclic groups, and finitely many others.

4.4 Uniqueness and Automorphisms

Those who study beautiful combinatorial structures know that often a strong uniqueness theorem can be used to show the existence of automorphisms. That is the case with the Shrikhande graph, which we use as an example. We will use the proof of Shrikhande's theorem (Theorem 3.8), which you may wish to review before reading what follows.

Theorem 4.7 *(a) Let G_1 and G_2 be copies of the Shrikhande graph, with vertices v_1 and v_2, respectively. Then there is an isomorphism from G_1 to G_2 which maps v_1 to v_2.*

(b) Let v be a vertex of the Shrikhande graph G. Then the group of automorphisms of G which fix v is the dihedral group of order 12.

Proof This follows because the uniqueness proof shows that the neighbourhood of a vertex is a hexagon H and the whole graph is built from H in a unique way. □

Corollary 4.8 *The Shrikhande graph is vertex-transitive and edge-transitive, and its automorphism group has order 192.*

Proof Take $G_1 = G_2 = G$ in part (a) of the theorem, which then asserts that there is an automorphism of G carrying v_1 to v_2. Now the edge-transitivity comes from the fact that the stabilizer of v acts transitively on the edges containing v, and the order of the automorphism group is a consequence of the Orbit-Stabilizer Theorem (Theorem 4.1). □

Exercises

4.1 Show that the triangular graphs $T(n)$ and the square lattice graphs $L_2(n)$ are rank 3 graphs.

4.2 Let $\text{Cay}(\Gamma, S)$ be a Cayley graph which has degree 3. Show that S consists of three involutions (elements of order 2) or it contains one involution and one inverse pair of elements of order greater than 2.

4.3 Let S be an inverse-closed subset of a group Γ. Suppose that the automorphism group of Γ contains a subgroup Δ such that S is an orbit of Δ. Show that $\text{Cay}(\Gamma, S)$ is edge-transitive.

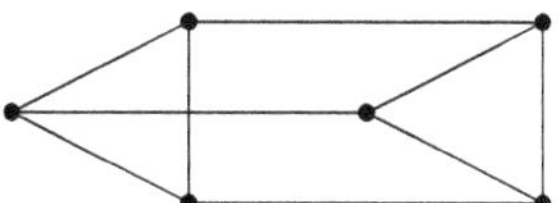

Figure 4.1 A triangular prism

Show that not every edge-transitive Cayley graph arises in this way. (Hint: consider the triangular prism (Fig. 4.1), and show that it is a Cayley graph $\mathrm{Cay}(C_6, \Delta)$, where Δ contains two elements of order 3 and one of order 2.)

4.4 A graph G is said to be 2-arc transitive if its automorphism group acts transitively on the set of all triples (v_1, v_2, v_3) of vertices such that $\{v_1, v_2\}$ and $\{v_2, v_3\}$ are edges and $v_1 \neq v_3$. Such a triple is called a *2-arc*.

(a) Prove that G is 2-arc transitive if and only if it is vertex-transitive and the stabilizer of a vertex v acts 2-transitively on the set $N(v)$ of neighbours of v.

(b) Deduce that if G is connected and 2-arc transitive, then either G is a complete graph or G contains no triangles.

(c) Show that the complete bipartite graph $K_{n,n}$, the 5-cycle, the Petersen graph and the Clebsch graph are all 2-arc transitive.

4.5 Let G be a vertex-transitive graph on n vertices. Show that the product of the clique number and the independence number of G is at most n.

Show that this assertion does not hold for all regular graphs.

5

Further Properties

In this chapter, we give the various properties and parameters of the Shrikhande graph; these have all been introduced earlier.

Vertices, edges, regularity The Shrikhande graph has 16 vertices and 48 edges. It is regular with degree 6. It is strongly regular: any two vertices have two common neighbours.

Symmetry The Shrikhande graph is vertex, edge and flag transitive. Its automorphism group is $(\mathbb{Z}_4)^2 : D_{12}$, of order 192, and has two orbits on non-edges.

Euler and Hamilton Since all vertices have even degree, the graph is Eulerian.

It is also Hamiltonian. Consider the description in Section 2.6. By using only four of the six types of edges given there (joining (x,y,z) to $(x,y+1,z-1)$, $(x,y-1,z+1)$, $(x+1,y,z-1)$ and $(x-1,y,z+1)$), we see that it contains $C_4 \square C_4$ as a spanning subgraph. Labelling the vertices by their x and y coordinates, we have a Hamiltonian cycle

$$(00,01,02,03,13,12,11,10,20,21,22,23,33,32,31,30).$$

By Dirac's Theorem 2.2, the complement of the Shrikhande graph is also Hamiltonian.

Cliques and independent sets The induced subgraph on the neighbourhood of any vertex is a 6-cycle; so the clique number of the Shrikhande graph is 3. This implies that the clique cover number is at least $\lceil 16/3 \rceil = 6$. We leave as an exercise the proof that the vertices can be covered with six cliques.

The set $\{(0,0,0),(0,2,2),(2,0,2),(2,2,0)\}$ is an independent set of size 4; there cannot be a larger one, so the independence number is 4. The chromatic number is also 4; for the 4-clique just shown is a subgroup, and its cosets give a 4-colouring.

The Shrikhande graph is not weakly perfect (its clique number and chromatic number are unequal). Hence it is not perfect. Indeed its complement is not weakly prefect either. (The Weak Perfect Graph Theorem of Lovász shows that the complement of a perfect graph is also perfect, and thus a perfect graph has clique cover number equal to its independence number. Alternatively, we can find an induced subgraph C_5; in terms of the description in Section 3.5, we can take four consecutive vertices a,b,c,d on the hexagon and the vertex ad joined to the vertices of the long diagonal.)

Since it is not perfect, it also does not belong to various important subclasses of perfect graphs, such as cographs and split graphs.

4-colourings It can be shown that the Shrikhande graph has 240 proper colourings with four colours. However, its chromatic polynomial is not known.

Eigenvalues and spanning trees The eigenvalues of the adjacency matrix of the Shrikhande graph are 6, 2 and -2, with multiplicities 1, 6 and 9. Since it is 6-regular, its Laplacian matrix $\mathscr{L}$ is given by $\mathscr{L} = 6I - A$, and so has eigenvalues 0, 4 and 8 with multiplicities 1, 6 and 9.

By the Matrix-Tree Theorem 2.11, the number of spanning trees is $4^6 \cdot 8^9/16 = 34359738368$.

Isoperimetric number Theorem 2.12 shows that the isoperimetric number of the Shrikhande graph is at least 2. On the other hand, there is a set S of 8 vertices inducing a subgraph of degree 4, so that $|\partial(S)| = 16$; thus the isoperimetric number is equal to 2. We invite readers to find such a set.

Distance-hereditary The Shrikhande graph is not distance-hereditary. Two vertices which are opposite in the 6-cycle induced by a vertex neighbourhood have distance 3 in the induced subgraph but 2 in the whole graph.

Open problem Find the chromatic polynomial of the Shrikhande graph.

Primality A graph G is *prime* (with respect to one of the graph products discussed in Fig. 2.9 if it is not isomorphic to the product of two smaller graphs.

Proposition 5.1 *The Shrikhande graph is prime with respect to the Cartesiain, categorical, strong or lexicographic product.*

Proof We give the proof for the Cartesian product here. The other cases we leave as exercises.

Let S be the Shrikhande graph, and suppose that S is isomorphic to $G \,\square\, H$, where G and H have more than one vertex, and $|V(G)| \leq |V(H)|$. Then either $|V(G)| = 2$ and $|V(H)| = 8$, or $|V(G)| = |V(H)| = 4$.

In the Cartesian product $G \,\square\, H$, the distance $d_{G\square H}((g,h)(g',h'))$ between two vertices (g,h) and (g',h') is given by

$$d_{G\times H}((g,h)(g',h')) = d_G(g,g') + d_H(h,h').$$

see Proposition 5.1.

This implies that a nontrivial Cartesian product $G \,\square\, H$ can have diameter 2 if and only if both factors have diameter 1, which is only possible if both factors are complete.

Hence, in the first case, S would be regular of degree 8, because in $G \,\square\, H$ the degree of a vertex (g,h) is $d_G(g) + d_H(h)$.

In the second case $G = H = K_4$. We already saw that $K_n \,\square\, K_n$ is isomorphic to the square lattice graph $L_2(n)$, and we know that $L_2(n)$ is not isomorphic to the Shrikhande graph.

We are grateful to Wilfried Imrich for this result.

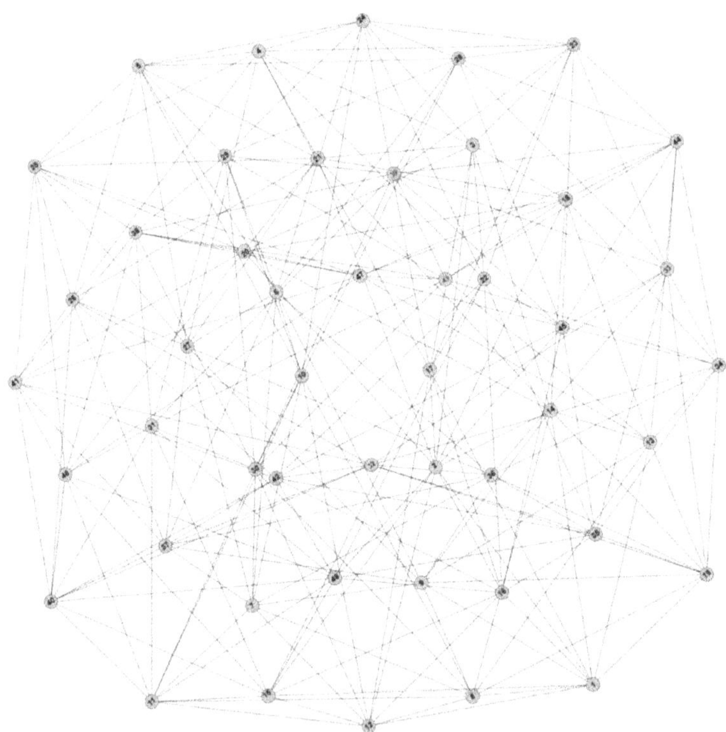

Figure 5.1 The line graph of the Shrikhande graph

Line graph The Shrikhande graph is not a line graph (we ask for a proof of this in the Exercises section).

For the record, we include in Fig. 5.1 a picture of the line graph of the Shrikhande graph.

Triangle graph A related graph is the *Dyck graph*, named after Walther von Dyck. It is the *triangle graph* of the Shrikhande graph: this means that its vertices are the triangles of the SG, two vertices joined in the Dyck graph if the corresponding triangles share an edge.

The Dyck graph has 32 vertices and is regular with degree 3, and has diameter 5 and girth 6. Its automorphism group is isomorphic to that of the Shrikhande graph, with order 192.

We will revisit the Dyck graph in Chapter 7, where, among other things, we will see that it is bipartite.

Exercises

5.1 Show that the vertex set of the Shrikhande graph can be covered with six cliques.

5.2 Find a set of eight vertices in the Shrikhande graph which induces a subgraph which is regular of degree 4.

5.3 Which of the forbidden subgraphs for the distance-hereditary property (Fig. 2.11) does the Shrikhande graph contain as induced subgraph?

5.4 Show that the Shrikhande graph is prime for each of the categorical, strong, and lexicographic products.

5.5 This exercise outlines a proof that the Shrikhande graph is not a line graph. Suppose that $SG \cong L(G)$ for some graph G.

    (a) Show that G is a graph with 16 edges, and is regular with degree 4.

    (b) Deduce that G has 8 vertices.

    (c) Show that if a graph has a vertex of degree k, then its line graph contains a complete subgraph on k vertices.

    (d) Deduce that SG is not a line graph.

5.6 Show that the line graph of the Shrikhande graph has 48 vertices and has valency 10. Deduce that it is Eulerian. Is it Hamiltonian? What is its automorphism group?

PART IV

The Shrikhande Graph in Context

6

Latin Squares

6.1 Introduction

Let n be a positive integer. A *Latin square* of order n is an $n \times n$ array whose entries are taken from an alphabet of size n, in such a way that each row or column of the array contains each letter in the alphabet precisely once.

The "letters" in the alphabet could be letters, numbers, colours or indeed any distinguishable symbols. Figure 6.1 gives two examples.

As far as we know, the first person to study Latin squares was the Korean mathematician Choi Seok-jeong (1646–1715), in his book *Gusuryak* (Fig. 6.2). The name was given by the Swiss mathematician Leonhard Euler (1707–83); we will see why he chose this name later.

We regard two Latin squares as being "essentially the same" (the technical term is *isotopic*)) if one can be changed into the other by rearranging rows and columns and/or changing the symbols in the alphabet. Up to isotopism, there is only one Latin square of each of the orders 1, 2 and 3, but there are two of order 4, as shown in Fig. 6.1.

Why are they different? The first square in Fig. 6.1 has the property that if you choose two rows and two columns such that the chosen entries in the first row are X and Y and the first chosen entry in the second row is Y, then the second chosen entry in this row will be always X. This property does not hold in the second square, as we can see by looking at the first two rows and the first

A	B	C	D
B	A	D	C
C	D	A	B
D	C	B	A

A	B	C	D
B	C	D	A
C	D	A	B
D	A	B	C

Figure 6.1 Two Latin squares of order 4

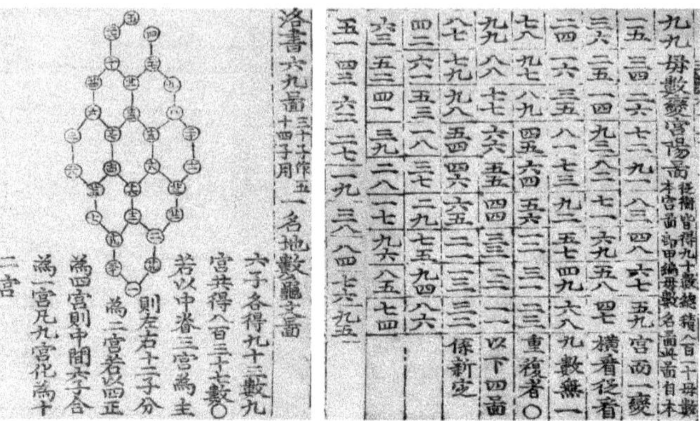

Figure 6.2 *Gusuryak* by Choi Seok-jeong

two columns. But the property is preserved by any rearrangement of the rows and columns and by any replacement of the alphabet symbols by new ones.

6.2 Euler and Latin Squares

Euler was originally interested in magic squares, a topic which has fascinated mathematicians of many different cultures since antiquity.

A *magic square* is an $n \times n$ array of numbers with the property that the sum of the entries in any row, or any column, or either of the two diagonals, is equal to a number called the *magic constant* of the square. Many constructions were found in olden times. Figure 6.3 shows Dürer's picture *Melancholia*, with the magic square shown by the side of the picture for clarity. The entries of the square are the positive integers from 1 to 16, and the magic constant is 34; Dürer has also arranged that the date of the picture is shown in the middle squares of the bottom row.

Choi was also interested in magic squares, and may have anticipated Euler's construction, which is described later.

Euler's construction can be expressed as follows. He defined a *Graeco-Latin square* to be an $n \times n$ array, each cell containing a pair of letters, one from each of two alphabets A and B (which Euler took to be the Latin and Greek alphabets) with the two properties

- the letters from each alphabet form a Latin square;
- given any pair of letters, one from each alphabet, there is exactly one cell of the array which contains this pair.

16	3	2	13
5	10	11	8
9	6	7	12
4	15	14	1

Figure 6.3 Dürer's *Melancholia*

$C\beta$	$A\gamma$	$B\alpha$
$A\alpha$	$B\beta$	$C\gamma$
$B\gamma$	$C\alpha$	$A\beta$

21	02	10
00	11	22
12	20	01

7	2	3
0	4	8
5	6	1

8	3	4
1	5	9
6	7	2

Figure 6.4 Euler's construction

Now take the letters in each alphabet to be $0,\ldots,n-1$ (the digits in the base n). Write a two-digit number in each cell: first the entry in A, then that in B. Each two-digit number in base n (from 0 to $n^2 - 1$) occurs once, and it is easily checked that all row and column sums are $n(n^2 - 1)/2$. To see this, note that in any row or column, each digit occurs once in the units position and once in the ns position; the sum of the digits is $n(n-1)/2$, so the sum of all the numbers is $(n+1)n(n-1)/2$.

This construction still has two minor drawbacks: first, it uses the numbers from 0 to $n^2 - 1$ rather than 1 to n^2; and second, we have not done anything about the diagonal sums. The first problem is easily fixed by simply adding one to each number in the square; this adds n to each row and column sum, so that they are now $n(n^2 + 1)/2$. The second requires a little trial and error, which we have already done (Fig. 6.4).

To express this in modern terminology, we make a definition. Two Latin squares L_1 and L_2 of order n, with alphabets A and B, are said to be *orthogonal* if, when they are superimposed on each other, each pair (a,b) consisting of a letter from A and a letter from B occurs exactly once in the resulting square. More generally, we say that a set $L_1,\ldots,L_k$ of Latin squares, all of the same order, is a set of *mutually orthogonal Latin squares*, or MOLS for short, if any two squares L_i and L_j are orthogonal.

So the question is as follows: for which n can two orthogonal Latin squares be constructed? Euler knew how to do this for all numbers n not congruent to 2 mod 4, and guessed that it was impossible for the remaining values. The case

Figure 6.5 Euler's 36 officers on parade

$n = 2$ is simple (try it yourself!). The case $n = 6$ was formulated by Euler as follows:

> *Six different regiments have six officers, each one holding a different rank (of six different ranks altogether). Can these 36 officers be arranged in a square formation so that each row and column contains one officer of each rank and one from each regiment?*

Trial and error suggests that the answer is "no" (Fig. 6.5).

This was not proved until 1900, and it was not until the 1960s that it was shown that for all larger numbers of this form ($n = 10, 14, \ldots$) there does exist a pair of orthogonal Latin squares: Euler was wrong! The three mathematicians who showed this (Bose, Shrikhande and Parker) were referred to as the "Euler spoilers". We will return to this later.

6.3 Constructing and Counting Latin Squares

How do we build a Latin square? To show that one exists for every order n, we can simply write n distinct symbols in the first row and then cycle them to produce the other rows: to get from any row to the next, move the first element to the end of the row and all the other elements one place to the left.

It is not obvious, but in fact if we have built any $k \times n$ "Latin rectangle" with $k < n$ (having k rows, n columns, n letters, so that each letter occurs once in each row and at most once in each column), we can complete it to a Latin square by adding extra rows. To show this, it suffices to show that it is always possible to add one more row, then continue adding rows until we reach a Latin square. That we can add a row follows from *Hall's marriage theorem*, to which we now turn.

Suppose that we are seeking husbands for n girls. In these enlightened times, we will require that each girl is to be married to a boy she knows. Is it possible? Clearly we require some conditions. Each girl must know at least one boy. Also, any two girls must between them know at least two boys, and more generally, any k girls must know at least k boys. Hall showed that, if this condition holds, then the choice of husbands is indeed possible. First, we make a definition.

Let $(A_1, A_2, \ldots, A_n)$ be a family of sets. A *system of distinct representatives* (or SDR, for short) for these sets is a family $(a_1, a_2, \ldots, a_n)$ of elements such that

- $a_i \in A_i$ for $i = 1, \ldots, n$ (that is, the elements are representatives of the sets);
- $a_i \neq a_j$ for $i \neq j$ (that is, the elements are distinct).

Clearly a solution to the marriage problem is an SDR for the sets $(A_1, \ldots, A_n)$, where A_i is the set of boys that the ith girl knows.

Theorem 6.1 *Let $(A_1, A_2, \ldots, A_n)$ be a family of sets. Then there is a system of distinct representatives for these sets if and only if*

$$\left| \bigcup_{i \in I} A_i \right| \geq |I|$$

for every set $I \subseteq \{1, 2, \ldots, n\}$.

Proof It is clear that the condition is necessary, since if $(a_1, a_2, \ldots, a_n)$ is an SDR then $\bigcup_{i \in I} A_i$ contains at least the elements a_i for $i \in I$.

The proof of sufficiency is a bit more complicated. Our proof is by induction on n. We denote $\bigcup_{i \in I} A_i$ by $A(I)$ for ease of writing, and call a set I of indices *critical* if $|A(I)| = |I|$. Now we separate two cases:

Case 1: There is no critical set of indices apart from possibly the whole set
$I = \{1, \ldots, n\}$. In this case, choose any representative a_n for A_n, and remove it from the other sets; that is, let $A_i' = A_i \setminus \{a_n\}$ for $i = 1, \ldots, n-1$. Since Hall's condition holds with at least one to spare for all subsets of $\{1, \ldots, n-1\}$, it continues to hold in the new family $(A_1', \ldots, A_{n-1}')$. By induction, these sets have an SDR, say $(a_1, \ldots, a_{n-1})$; adding in a_n as representative for A_n gives the required SDR for the original family.

Case 2: There exist critical proper subsets of $\{1, \ldots, n\}$. Let J be a maximal such that $|A(J)| = |J|$. By induction we can choose an SDR for the sets A_j for $j \in J$. Now for $i \notin J$, let $A_i^* = A_i \setminus A(J)$. A fairly short argument shows that the subsets $(A_i^* : i \notin J)$ satisfy Hall's condition, and so have an SDR. Putting this together with the SDR for the remaining sets gives the required SDR for the entire family.

We apply this result by means of a simple but useful consequence.

Corollary 6.2 *Let $(A_1, A_2, \ldots, A_n)$ be a family of subsets of a set X. Suppose that for some $r > 0$,*

(a) $|A_i| = r$ for $i = 1, 2, \ldots, n$;
(b) any point of X lies in exactly r of the sets $A_1, \ldots, A_n$.

Then $(A_1, A_2, \ldots, A_r)$ has a system of distinct representatives.

Proof Take any set I of indices, and let $Y = \bigcup_{i \in I} A_i$. Count pairs (i, y) with $i \in I$ and $y \in Y$. For each i there are r choices of y. Also, each y lies in r sets A_i, but these may not all have index in I; so the number of choices of $i \in I$ is at most r. We conclude that $|Y|r \geq |I|r$, so $|Y| \geq |I|$. Thus Hall's condition is satisfied, and there is an SDR.

Now we can show:

Theorem 6.3 *If $k < n$, then any $k \times n$ Latin rectangle can be extended to an $(k + 1) \times n$ rectangle by adding a row, and hence can be extended to an $n \times n$ Latin square.*

Proof Let R be the rectangle, and set $r = n - k$. Let A_i be the set of letters not occurring in the ith column of R. Then $|A_i| = n - k = r$. Also, each letter x occurs k times in R, in k distinct columns, so lies in $r = n - k$ of the sets A_i.

By the preceding corollary, the family $(A_1, \ldots, A_n)$ has an SDR, say $(a_1, \ldots, a_n)$. Now we can add this n-tuple as the next row to produce a larger Latin rectangle.

These arguments can be tweaked to show more. We will not spell these out in detail, but point readers to other accounts.

In Corollary 6.2, it is possible to show that there exist at least $r!$ distinct SDRs for any $k \times n$ Latin rectangle with $r = n - k > 0$. This gives us a lower bound on the number of $n \times n$ Latin squares: the numbers of ways of choosing the first, second, $\ldots$, nth row are at least $n!$, $(n-1)!$, $\ldots$, $1!$ respectively. So the number of Latin squares is at least

$$n! \cdot (n - 1)! \cdots 1!,$$

a rather large number!

But it is known that even this is an underestimate for the total number of Latin squares. For example, the first row is an arbitrary permutation, and there are indeed $n!$ possibilities. Without loss of generality, we can suppose that the first row is $(1, 2, \ldots, n)$. Then the second row $(a_1, a_2, \ldots, a_n)$ is a permutation satisfying $a_i \neq i$ for all i; thus it is a *derangement*. According to one of the oldest results in enumerative combinatorics, the number of derangements is the integer closest to $e^{-1} n!$, where e is the base of natural logarithms; this is strictly greater than $(n - 1)!$ for all $n > 3$.

Indeed, the number of Latin squares of order n is known to be bounded above and below by $(cn)^{n^2}$ (with possibly different constants c in the two bounds). We refer to [71] for an account of this.

6.4 The Euler Conjecture

6.4.1 Rise

How many squares can there be in a set of mutually orthogonal Latin squares of order n? (We abbreviate "mutually orthogonal Latin squares" to MOLS.)

Proposition 6.4 *The number of mutually orthogonal Latin squares of order n is at most $n - 1$.*

Proof Suppose that $L_1, L_2, \ldots, L_r$ are Latin squares of order n, any two of which are orthogonal.

We can change the symbols in each square so that its first row contains the numbers $1, 2, \ldots, n$ in order. (Applying any one-to-one map to the symbols in any square does not change its orthogonality to the others.) Now consider the entries in the second row and first column in the squares. None of these entries can be 1, since each square already has 1 in the first column. Also, no two entries in this position in different squares can be equal. For given any two squares, say L_i and L_j, they already both have the entry k in the $(1, k)$ position; so this cannot also occur in the $(2, 1)$ position, by orthogonality. Since there are only $n - 1$ possible entries, there cannot be more than $n - 1$ squares.

A set of $n - 1$ mutually orthogonal Latin squares of order n is said to be *complete*. The existence of a complete set implies the existence of certain finite geometries; we will briefly look at this later.

Now complete sets exist if n is a power of a prime number:

Proposition 6.5 *Suppose that $n = p^a$, where p is prime and $a \geq 1$. Then there exists a complete set of MOLS of order n.*

Proof Our proof here relies on an important discovery of Évariste Galois, the French mathematician who was killed in a duel in 1832 at the age of 19, having

+	0	1	α	β
0	0	1	α	β
1	1	0	β	α
α	α	β	0	1
β	β	α	1	0

$\times$	0	1	α	β
0	0	0	0	0
1	0	1	α	β
α	0	α	β	1
β	0	β	1	α

Figure 6.6 Addition and multiplication tables for a 4-element field

already laid the foundations of group theory. Among his discoveries published in his lifetime was the following:

There exists a finite field of order n if and only if n is a prime power.

A *field* is a structure with two operations, addition and multiplication, satisfying the usual laws of arithmetic: the commutative and associative laws for both addition and multiplication, the distributive law, the existence of elements 0 and 1, which are identities for addition and multiplication respectively, and the existence of additive inverses for all elements and multiplicative inverses for all non-zero elements. Familiar number systems such as the rational, real and complex numbers are fields. The best known finite fields are given by the integers modulo p, where p is a prime number. Warning: the field in Galois' result is *not* the integers modulo n. We give addition and multiplication tables for the Galois field with four elements $0, 1, \alpha, \beta$ (Fig. 6.6).

Using a finite field, the construction of the MOLS is easy. Let

$$F = \{a_0 = 0, a_1, a_2, \ldots, a_{n-1}\}$$

be a Galois field of order $n = p^e$. Then define squares $L_1, \ldots, L_{n-1}$, each with rows and columns indexed by F, with the entry in row x and column y of square L_i equal to $a_i x + y$. Now it is straightforward to show that each is a Latin square and that any two are orthogonal.

Proposition 6.6 *Let n, m, k be positive integers, and suppose that there exist k MOLS of order n and k MOLS of order m. Then there exist k MOLS of order mn.*

Proof Let L and L' be Latin squares of orders n and m respectively; we may suppose that the rows and columns of L are indexed by $\{1, 2, \ldots, n\}$ and that these are the symbols of L, with a similar statement for L'. Then we construct a Latin square denoted by $L \otimes L'$, whose rows and columns are indexed by the ordered pairs (x, x') with $1 \le x \le n$ and $1 \le x' \le m$; these ordered pairs will also be the symbols of the square. If L has (i, j) entry k, and L' has (i', j') entry k', then we put entry (k, k') in row (i, i') and column (j, j') in $L \otimes L'$.

Now it can be verified that $L \otimes L'$ is a Latin square. Moreover, if L_1 and L_2 are orthogonal Latin squares of order n, and L'_1 and L'_2 are orthogonal Latin squares of order m, then $L_1 \otimes L'_1$ and $L_2 \otimes L'_2$ are orthogonal Latin squares of order nm.

So the result of the proposition follows.

From these results we deduce *McNeish's Theorem*:

Theorem 6.7 *Let* $n = p_1^{e_1} p_2^{e_2} \cdots p_r^{e_r}$, *where* $p_1, \ldots, p_r$ *are distinct primes and the exponents* $e_1, \ldots, e_r$ *are positive. Let*

$$q = \min\{p_1^{e_1}, p_2^{e_2}, \ldots, p_r^{e_r}\}.$$

Then there exist $q - 1$ *MOLS of order* n.

Proof By assumption and Proposition 6.5, there exist $q - 1$ MOLS of order $p_i^{e_i}$ for $i = 1, 2, \ldots, r$; so by Proposition 6.6, the result follows.

Now we can establish the fact known to Euler (though his proof was different):

Corollary 6.8 *If* n *is not congruent to* 2 *modulo* 4, *then there exist two orthogonal Latin squares of order* n.

Proof If $n \not\equiv 2 \pmod 4$, then the smallest prime power in the factorization of n is at least 3, so the number of MOLS is at least 2.

Euler conjectured that the converse was also true; namely, if n is congruent to 2 modulo 4, then orthogonal Latin squares of order n do not exist.

McNeish then extended Euler's conjecture to the bolder conjecture that the lower bound in Theorem 6.7 is exact; that is, given the hypotheses of the theorem, the maximum number of MOLS of order n is $q - 1$.

The first indication that these conjectures may be false came with the construction of a set of three MOLS of order 21 by E. T. Parker. (McNeish's conjecture would give 2 for the maximum number.) This led on to the work of the Euler spoilers (Fig. 6.7), to which we turn in the next subsection.

6.4.2 Fall

We noted earlier that Parker showed the falsity of NcNeish's conjecture. This suggested that perhaps Euler's long-standing conjecture might also be false; Bose and Shrikhande set to work and found a pair of orthogonal Latin squares of order 22, thereby disproving the conjecture. Parker then used computer search to find a pair of orthogonal Latin squares of order 10. Finally, the three authors pooled their resources and completely destroyed the conjecture:

Figure 6.7 The Euler spoilers in the *New York Times*

Theorem 6.9 *For every integer n > 1, except for n = 2 and n = 6, there exists a pair of orthogonal Latin squares of order n.*

The demise of a centuries-old conjecture featured on the front page of the *New York Times*. An editorial in that paper on 27 April 1959 stated that it would be a serious mistake to suppose that modern mathematics was far from real life. Actually, there had never been a time in history when mathematics was so widely applied in so many different fields for so many vital purposes, as was true then. What was the case in 1959 is still true today!

A proof of the theorem would take us well beyond our story; we simply give the construction for order 22, found by Bose and Shrikhande [22].

In 1850, Thomas Kirkman, vicar at Croft in Lancashire, England, and part-time mathematician, published a solution [64] to the following problem he had posed in the *Lady's and Gentleman's Diary* in 1844:

> Fifteen young ladies of a school walk out three abreast for seven days in succession: it is required to arrange them daily so that no two shall walk abreast more than once.

We will not detour to give a solution here. It is possible to use trial and error (as Kirkman did) or computation, but the simplest way is to use ideas from finite geometry; we refer to Primrose [80] for an account of this.

We also need a finite field construction. We have seen that, if q is a prime power, then there is a finite field of order q, from which we can construct $q - 1$ mutually orthogonal Latin squares of order q. We need Latin squares with an extra property, and we compromise by finding one fewer.

Let L be a Latin square of order n, with entries $1, 2, \ldots, n$. We say that L is *idempotent* if the entry in row i and column i is i, for each $i \in \{1, \ldots, n\}$.

(If L is the multiplication table of an algebraic structure, this says that each element of the structure is idempotent, that is, equal to its square.) A similar definition works if the rows and columns of L are indexed by an arbitrary set I, and the letters in L are taken from I.

Theorem 6.10 *Suppose that F is a finite field of order q. Then there exist $q - 2$ mutually orthogonal idempotent Latin squares of order q.*

Proof We index the Latin squares by elements $t \in F \setminus \{0.1\}$; the entry in row x and column y of L_t is $tx + (1 - t)y$. It is easy to check that L_t is a Latin square, and is idempotent. To show that L_t and L_s are orthogonal for $s \neq t$, we have to show that the equations

$$tx + (1 - t)y = a, \qquad sx + (1 - s)y = b$$

have a unique solution (x, y) for given a, b; but the determinant of these two equations is

$$\begin{vmatrix} t & 1 - t \\ s & 1 - s \end{vmatrix} = t - s \neq 0,$$

so the result follows. $\square$

Now Bose and Shrikhande proceed as follows. The 22 coordinates indexing the rows and columns of the Latin squares will be given by the set S of 15 schoolgirls and the set D of 7 days of the week. Now consider the following collection of subsets of $S \cup D$:

(a) for each day $d \in D$, and each triple T of schoolgirls walking together on
     day d, the set $\{d\} \cup T$, of size 4;
(b) the set D, of size 7.

Let $\mathscr{B}$ be this collection of sets, which we will call *blocks*.

Now observe that any two elements of $S \cup D$ are contained in a unique block. For a pair of schoolgirls, this is the condition of the problem; for a schoolgirl and a day, it is clear; and for two days, the set D is the one required. A structure like this is called a *pairwise balanced design*.

The cardinalities of the blocks are 4 and 7, both prime powers. So for each block $B \in \mathscr{B}$, we construct a pair of orthogonal idempotent Latin squares (as in Theorem 6.10), where the rows and columns are indexed by the elements of B which also appear as the entries in both squares. (To do this, choose a bijection from the finite field of order $|B|$ to B, and replace the row and column indices and entries in the theorem by their images under this bijection.) Call these squares $L_1(B)$ and $L_2(B)$.

Now we construct two Latin squares L_1 and L_2 of order 22 as follows. Rows and columns are indexed by the set of points of the design (schoolgirls and days). In row x and column x, we put the entry x in both squares. If $x \neq y$, there is a unique block B containing x and y; take the (x,y) entries of L_1 and L_2 to be the (x,y) entries of $L_1(B)$ and $L_2(B)$.

It is routine to check that these are orthogonal Latin squares.

Note, incidentally, that they are both idempotent. This opens the door to the use of recursion, which was used by the Euler spoilers in further constructions, and has been refined and extended by many other authors, notably Richard Wilson, for combinatorial constructions.

Very soon afterwards, Chowla, Erdős and Strauss [37] applied a number-theoretic argument to the Euler spoilers' argument and obtained the following stronger result:

Theorem 6.11 *There is a function f, such that $f(n) \to \infty$ as $n \to \infty$, such that there exist at least $f(n)$ mutually orthogonal Latin squares of order n.*

Chowla *et al.* gave $f(n) \geq \frac{1}{3}n^{1/91}$; this was soon improved by Wilson [108] to $f(n) \geq n^{1/17} - 2$. The current best bound is about $n^{1/14.2}$.

So, contrary to McNeish's guess, there is a lower bound for the number of MOLS, tending to infinity with the order n, and not depending on the prime power factorization of n. Thus, for any r, there are only finitely many integers n for which a set of r MOLS of order n does not exist.

A curious footnote to Euler's conjecture has been added recently: Rather *et al.* [86] showed that there do exist two quantum orthogonal Latin squares of order 6. But it would take us too far afield to explain what this means.

6.4.3 Finite geometries

There is a close connection between complete sets of MOLS and finite geometries, which we now describe.

Begin with a complete set $L_1, L_2, \ldots, L_{n-1}$ of MOLS of order n. From it, we can construct a geometry of points and lines as follows:

- the points are the ordered pairs (i,j), for $1 \leq i,j \leq n$ (these correspond to the points of the $n \times n$ square grid);
- the lines are of two types:
  - the horizontal and vertical lines of the grid (the sets of pairs (i,j) with fixed j and arbitrary i, or fixed i and arbitrary j);
  - for each k with $1 \leq k \leq n - 1$, and each symbol a of the Latin square L_k, the set of positions in the grid where the symbol a occurs in L_k.

There are n^2 points and $n(n + 1)$ lines ($2n$ of the first type and $n(n - 1)$ of the second); each line contains n points. We observe two further properties:

(a) two points lie on exactly one line;
(b) if we call two lines *parallel* if they are equal or disjoint, then parallelism is an equivalence relation, and the lines fall into $n + 1$ parallel classes, each class containing n lines covering all the points.

For (a), take distinct points (i_1, j_1) and (i_2, j_2). If $j_1 = j_2$, they lie on a horizontal line, and if $i_1 = i_2$ they lie on a vertical line. Suppose that neither inequality holds. Since the set of Latin squares is complete, there is a unique square L_k which has the same symbol a in the two positions; so the line consisting of the positions where a appears in L_k contains both points.

For (b), the horizontal and vertical lines form two parallel classes, and the lines defined by the positions of the symbols in each square L_k form a parallel class for each k.

These two properties remind us of the points and lines of the familiar Euclidean plane; a geometry satisfying them is called an *affine plane*. In particular, the geometry just constructed is an affine plane of *order n*.

Conversely, given an affine plane of order n, we can reconstruct a set of $n-1$ MOLS as follows. Choose two of the parallel classes to be the horizontal and vertical lines; this identifies the point set with the $n \times n$ grid. Then, for each further parallel class, associate one symbol with each line of the class, and put that symbol in the positions corresponding to points on that line. Then it can be checked that the $n - 1$ arrays constructed are Latin squares, and that any two are mutually orthogonal.

It is possible to go a little further, to construct a projective plane, following ideas from the theory of perspective developed by artists in the Renaissance. We add to the affine plane a "line at infinity" as follows: for each parallel class of lines, add a new symbol, a "point at infinity" lying on each line of the parallel class (so $n + 1$ points altogether); then add one new line containing all the points at infinity. Now check that the resulting structure has the following properties:

- there are $n^2 + n + 1$ points and $n^2 + n + 1$ lines;
- each line contains $n + 1$ points, and each point lies on $n + 1$ lines;
- two points lie on just one line, and two lines intersect at just one point.

Such a structure is called a *projective plane* of order n.

Conversely, given a projective plane of order n, select one line, and delete it together with all of its points, and we obtain an affine plane of order n. Thus we have:

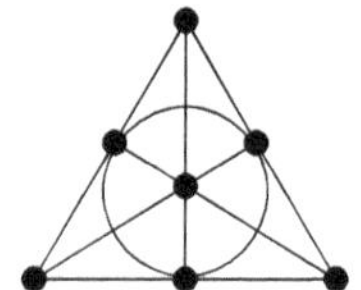

Figure 6.8 The Fano plane

Theorem 6.12 *For an integer $n > 1$, the following are equivalent:*

(a) there exists a complete set of MOLS of order n;
(b) there exists an affine plane of order n;
(c) there exists a projective plane of order n.

A hard unsolved problem in finite geometry asks: for which n are these conditions satisfied? The complete state of knowledge to date is summarised in the following theorem:

Theorem 6.13 *(a) If n is a prime power, there is a projective plane of order n.*
(b) If n is congruent to 1 or 2 (modulo 4), and n is not the sum of two squares, then there is no projective plane of order n.
(c) There is no projective plane of order 10.

We have seen the reason for (a): from the Galois field of prime power order, we can construct a complete set of MOLS. Part (b) is the celebrated *Bruck–Ryser Theorem* [27], while (c) is the result of a large computation by Lam and his collaborators [70]. At present, all known finite projective planes have prime power order.

Figure 6.8 shows the smallest projective plane, over the binary field of integers modulo 2. This is known as the *Fano plane*. The dots are the seven points, and the straight and curved lines are the seven lines.

6.5 Latin Square Graphs

From any Latin square, we can construct a strongly regular graph as follows.

The vertices of the graph are the n^2 cells of the square. We join two vertices if the cells lie in the same row or same column of the array or have the same entry.

Theorem 6.14 *The Latin square graph built from a Latin square of order n is strongly regular with parameters $(n^2, 3(n-1), n, 6)$.*

Proof The graph has valency $3(n-1)$, since there are $n-1$ cells in the same row, $n-1$ in the same column and $n-1$ with the same entry, and no overlaps among these.

Suppose that two cells lie in the same row (the other cases are similar). Suppose that the row number is i, the column numbers are j and k and the entries are a and b. Then there are $n-2$ further cells in row i: one in column j with entry b, and one in column k with entry a, so n altogether.

Now consider two cells which lie in different rows and different columns and have different entries. They have six common neighbours: the cell in the same row as the first and column as the second; same row as the first and entry as the second; ...; same entry as the first and column as the second.

We have defined two Latin squares to be isotopic if one can be obtained from the other by permuting rows and columns and changing the letter in the alphabet. There are some additional transformations that can be applied, involving changing the categories "row", "column" and "letter". For example, interchanging rows and columns has the effect of transposing the square. Assuming that the alphabet is $\{1, 2, \ldots, n\}$, interchanging rows and letters, changes the array with (i,j) entry k into the array with (k,j) entry i (which is also a Latin square). We call two Latin squares *paratopic* if one can be obtained from the other by a combination of an isotopism and a transformation permuting rows, columns and letters, as just described.

Theorem 6.15 *If $n > 4$, a Latin square can be reconstructed from its Latin square graph uniquely up to paratopism.*

Proof The Latin square graph has $3n$ "grand cliques" of size n, corresponding to the rows, columns and letters in the array. Any other clique in the graph has size at most 4. Suppose that one vertex of such a clique lies in row i, column j and has letter k. Any other vertex must be in row i, or in column j, or have letter k. Since at least two rows, columns and letters are involved, there are at most four cells, the extreme case occurring when two rows and two columns intersect in four cells containing just two letters. This forms a Latin subsquare of order 2, which is called an *intercalate*.

So, if $n > 4$, we can identify the three families of grand cliques. Using one family as rows and a second as columns gives an $n \times n$ square array, and the third family tells us how to put letters into the array.

6.6 Pseudo and Negative Latin Square Graphs

There are Latin square graphs constructed from sets of MOLS in a similar way. Suppose that $L_1, L_2, \ldots, L_{s-2}$ are mutually orthogonal Latin squares of order n. Then we make a graph whose vertices are the cells of the $m \times m$ square; two vertices are joined if one of the following holds:

- they lie in the same row;
- they lie in the same column;
- they contain the same symbol in one of the squares $L_1, \ldots, L_{s-2}$.

Theorem 6.16 *The graph just constructed is strongly regular, with parameters*

$$n = m^2, \quad k = s(m-1), \quad \lambda = s^2 - 3s + m, \quad \mu = s(s-1).$$

Proof We will not give a complete proof; the argument is similar to that we used for the earlier Latin square graphs.

There are clearly m^2 vertices. Given any vertex, there are $m-1$ vertices in the same row, $m-1$ in the same column and $m-1$ having the same symbol in each one of the squares, giving $s(m-1)$ altogether.

If v and w are joined, then suppose they lie in the same row; there are $m-2$ further vertices in that row, and for each pair of (columns, symbols of L_1, ..., symbols of L_{s-2}), there is one vertex joined to v for the first reason and to w for the second, giving $\lambda = (m-2) + (s-1)(s-2)$.

We leave counting the common neighbours of two non-adjacent vertices to the diligent reader. $\qquad\square$

The converse of this theorem is false. So a graph which has the parameters $(m^2, s(m-1), s^2 - 3s + m, s(s-1))$ is called a *pseudo-Latin square graph*, denoted by $PL_s(m)$; and if it does come from a family of MOLS, it is a *Latin square graph* $L_s(n)$. Thus what we previously called Latin square graphs are $L_3(n)$, and the notation $L_2(n)$ agrees with our previous usage.

One of the more curious stories in this corner of the subject concerns the work of Dale Mesner, a student of R. C. Bose. Noting that the parameters of the graph associated with a set of $s-2$ mutually orthogonal Latin square graphs of order m are

$$n = m^2, \quad k = s(m-1), \quad \lambda = s^2 - 3s + m, \quad \mu = s(s-1),$$

he observed that if we replace m and s by their negatives, we obtain

$$n = m^2, \quad k = s(m+1), \quad \lambda = s^2 + 3s - m, \quad \mu = s(s+1),$$

which are feasible parameters of a strongly regular graph, at least if $\lambda \geq 0$, that is, $s(s+3) \geq m$. He defined a strongly regular graph with these parameters to have *negative Latin square type*.

Clearly there is something special about the case $m = s(s+3)$, where this bound is met, and the graph has $\lambda = 0$, that is, no triangles. We note that the Clebsch graph is such an extremal graph of negative Latin square type, with $s = 1$ and $m = 4$.

Mesner started working on the second case, with $s = 2$ and $m = 10$, a strongly regular graph with parameters $(100, 22, 0, 6)$. After considerable effort, he was able to construct such a graph in his thesis in 1956, and to prove its uniqueness a few years later.

In 1968, the group theorists Donald Higman and Charles Sims, unaware of Mesner's work, gave a simple construction of the graph and proof that its automorphism group has a simple subgroup of index 2, the then-new sporadic simple group now known as the Higman–Sims group. As a result, group theorists refer to the graph as the *Higman–Sims graph*. It would be reasonable to call the graph the *Mesner graph*. It is a very interesting graph, one of only seven known strongly regular graphs containing no triangles apart from the complete bipartite graphs, and indeed it contains the other six as induced subgraphs.

Dale Mesner was philosophical about his failure to find a sporadic simple group, admitting that he knew little group theory.

Here is the construction by Higman and Sims. It uses the *Steiner system* $S(3, 6, 22)$ constructed by Witt in 1930: this is a configuration of 77 6-element subsets (called "blocks") of a set of 22 points, with the property that any three points lie in a unique block. Now take the vertex set of the graph to be $\{*\} \cup P \cup B$, where P and B are the sets of points and blocks of the Witt system and $*$ a new point. The edges are

- $\{*, p\}$ for all $p \in P$;
- $\{p, b\}$ for all $p \in P$, $b \in B$ with $p \in b$;
- $\{b, b'\}$ for all $b, b' \in B$ with $b \cap b' = \emptyset$.

For the record, the "magnificent seven" triangle-free strongly regular graphs known (with their parameters) are:

- the pentagon, or 5-cycle $(5, 2, 0, 1)$;
- the Petersen graph $(10, 3, 0, 1)$;
- the Clebsch graph $(16, 5, 0, 2)$;
- the Hoffman–Singleton graph $(50, 7, 0, 1)$;
- the Gewirtz graph $(56, 10, 0, 2)$;
- the 77-graph $(77, 16, 0, 4)$;
- the Mesner or Higman–Sims graph $(100, 22, 0, 6)$.

The 77-graph and Gewirtz graph are the induced subgraphs on the sets of non-neighbours of a vertex and an edge respectively in the Mesner graph. The Mesner graph can be partitioned into two sets, each inducing a copy of the Hoffman–Singleton graph.

More details on the work of Dale Mesner, properties of the graph and further developments can be found in the very rich paper by Klin and Woldar [65].

One curious footnote to this story is the fact that Higman and Sims constructed their graph while they were at a conference in Oxford. At the conference, Marshall Hall had described his construction, with David Wales, of a graph on 100 vertices which contained Janko's second sporadic simple group in its automorphism group. This computer construction was the first existence proof for Janko's group. At the conference dinner that evening, Higman and Sims discussed this and wondered whether there was another graph on 100 vertices whose automorphism group contained a new simple group; after dinner, they succeeded (by hand) in the construction outlined here. Curiously, the graph constructed by Hall and Wales was strongly regular, with parameters $(100, 36, 14, 12)$; that is, it is a pseudo-Latin square graph $PL_4(10)$ (but not a Latin square graph), while Higman and Sims (and Mesner) constructed a negative Latin square graph $NL_2(10)$.

6.7 $L_2(4)$ and the Shrikhande Graph

In this section, we return to the Shrikhande graph. We will see a remarkable connection between this graph and the question of whether a set of mutually orthogonal Latin squares can be extended to a larger set, and give a very simple construction of the Shrikhande graph: it is the complement of the Latin square graph of the Cayley table of the cyclic group of order 4.

Recall that a set of $s - 2$ MOLS of order n gives a strongly regular graph with the parameters

$$n = m^2, \quad k = s(m - 1), \quad \lambda = s^2 - 3s + m, \quad \mu = s(s - 1).$$

Not every graph with such parameters arises from Latin squares in this way. A graph with these parameters is called a pseudo-Latin square graph, denoted by $PL_s(m)$. It is a *Latin square graph $L_s(m)$* if it does come from a set of MOLS.

For $s = 1$, the parameters of $PL(1, m)$ have $\mu = 0$; so a $PL_1(m)$ graph is the disjoint union of m complete graphs of size m.

For $s = 2$, the parameters $(m^2, 2(m - 1), m - 1, 2)$ are those of Shrikhande's theorem, which can thus be stated in the form:

A pseudo-Latin square graph $PL_2(m)$ is a Latin square graph if $m \neq 4$; for $m = 4$, there is a unique exception.

To examine this question, consider the complement of $PL(s, m)$. Using the formulae for the parameters of the complement of a strongly regular graph given earlier, we see that the complement is a $PL(m + 1 - s, m)$.

Now suppose that we have a set of $s-2$ MOLS of order m. The complement of the Latin square graph $L_s(m)$ is a $PL(m+1-s,m)$; if it is a $L_{m+1-s}(m)$, then each Latin square is orthogonal to the given set, and they are orthogonal to one another, so we obtain altogether a set of $(s-2)+(m+1-s)=m-1$ MOLS of order m, that is, a complete set. We have proved:

Theorem 6.17 *Suppose that every pseudo-Latin square graph $PL_{m+1-s}(m)$ is a Latin square graph $L_{m+1-s}(m)$. Then any set of $s-2$ MOLS of order m can be extended to a complete set of $m-1$ MOLS of order m.*

We give a couple of corollaries. Let us say the *defect* of a set of MOLS is the number by which it falls short of a complete set.

Theorem 6.18 *(a) A set of MOLS of order m and defect 1 can be completed to a complete set.*
(b) A set of MOLS of order m and defect 2 can be extended to a complete set if $m \neq 4$.

What about $m=4$? In this case a set of defect 2 consists of a single Latin square. Each has a Latin square graph. Their complements are $PL_2(4)$ graphs; we know that there are two such graphs: $L_2(4)$ and the Shrikhande graph. We know that one of the Latin squares of order 4 (the Cayley table of the Klein group $\mathbb{Z}_2 \times \mathbb{Z}_2$) can be extended to a complete set of three MOLS, while the other one (the Cayley table of the cyclic group $\mathbb{Z}_4$) cannot be. (The two Latin squares are shown in Fig. 6.1.)

We also see that there is a simple description of the two strongly regular graphs with parameters $(16,6,2,2)$ occurring in Shrikhande's theorem. Take the two Latin squares of order 4, as described earlier. Form their Latin square graphs, which are strongly regular with parameters $(16,9,4,6)$. Take their complements: these are strongly regular with parameters $(16,6,2,2)$. One is $L_2(4)$; the other is the Shrikhande graph!

In fact, Shrikhande's theorem has been generalized in a different way, by the following result proved in consecutive papers by Bruck and Bose in the *Pacific Journal of Mathematics* in 1963 [20, 26]:

Theorem 6.19 *There is a function f on the natural numbers, tending to infinity with n, such that*

(a) a pseudo-Latin square graph $PL_s(n)$ with $s \le f(n)$ is a Latin square graph $L_s(n)$;
(b) a set of at least $n-f(n)-1$ MOLS of order n can be extended to a complete set of MOLS.

6.8 Automorphisms of Latin Square Graphs

In this section, we give a different approach to computing the automorphism group of the Shrikhande graph and $L_2(4)$, using Latin squares.

There are two groups of order 4: the cyclic group C_4 and the *Klein group* $C_2 \times C_2$. Their Cayley tables are Latin squares of order 4, so their Latin square graphs are strongly regular, with parameters $(16,9,4,6)$. By Proposition 3.2, their complements are strongly regular, with parameters $(16,6,2,2)$. By Shrikhande's theorem, they must be $L_2(4)$ and the Shrikhande graph. It turns out that the Klein group gives $L_2(4)$, while the cyclic group gives the Shrikhande graph.

Theorem 6.20 *The automorphism group of the Shrikhande graph has structure $(C_4 \times C_4) : (C_2 \times S_3)$, that is, the semidirect product of a normal subgroup of order 16 which is the direct product of two cyclic groups of order 4 and a subgroup of order 12 which is the direct product of the cyclic group of order 2 and the symmetric group of degree 3.*

Proof Since a graph and its complement have the same automorphism group, we will instead find the automorphism group of the complement of the Shrikhande graph, which as we saw is the Latin square graph associated with the Cayley table of the cyclic group of order 4.

For reference, the Latin square is shown in Fig. 6.9.

The group of order 16 is generated by the cyclic permutation of the four rows and the cyclic permutation of the four columns.

The group C_4 has just one non-trivial automorphism: the map $x \mapsto -x$ (mod 4). This automorphism, acting element-wise on the Cayley table, induces a graph automorphism.

Finally, there is a group of so-called parastrophisms, which permutes the sets of rows, columns and symbols among themselves.

These generate the group of order 192 in the theorem.

Finally, we have to show that the automorphisms listed form the full automorphism group. We do this by considering the structure of the Shrikhande graph established in our first proof of Shrikhande's Theorem. The induced

0	1	2	3
1	2	3	0
2	3	0	1
3	0	1	2

Figure 6.9 Cayley table of integers mod 4

subgraph on the neighbourhood of a vertex is a 6-cycle, and its automorphism group is the dihedral group $D_6 \cong C_2 \times S_3$. Now the proof of the theorem shows that an automorphism fixing all vertices in the neighbourhood of one vertex must be the identity. So the order of the subgroup fixing a vertex is at most 12, and we conclude that the full automorphism group has order at most $12 \cdot 16 = 192$. Thus we have found all the automorphisms.

In fact, a similar result applies to almost all Latin square graphs derived from group Cayley tables. If we start with a group G of order n, with $n > 4$, and produce the Latin square graph of the Cayley table of G, its automorphism group can be described in similar terms; it has a normal subgroup, which is a qoutient of $G \times G \times G$ by its centre, and the quotient is the direct product of the outer automorphism group of G and the symmetric group of degree 3. This is described in greater detail and proved in [3].

An essential step in the proof is the observation that a Latin square graph associated with a Latin square of order greater than 4 has just $3n$ cliques of size n, corresponding to the rows, columns and letters of the Latin square; so an automorphism must permute these objects among themselves.

This is false for $n = 4$. An *intercalate* in a Latin square is a subsquare of order 2, that is, a set of two rows and two columns whose four cells contain just two symbols. The cells of an intercalate form a clique of size 4. So a Latin square of order 4 containing intercalates has additional cliques of size 4. If we examine carefully the two Latin squares, we find that in the Cayley table of the cyclic group, although there are additional 4-cliques, they do not give rise to a different set of 12 such cliques falling into 3 classes of 4 pairwise disjoint cliques, so they do not contribute extra automorphisms. However, in the other case, there are 12 intercalates, and they can be partitioned into three sets of four pairwise disjoint cliques. So there is an extra automorphism, interchanging the two systems of 12 cliques, and the full automorphism group is twice as large as expected, namely 1152.

Figure 6.10 shows the 12 intercalates in the Cayley table of $C_2 \times C_2$. The top row shows the Latin square, and a labelling of the 16 positions. The bottom row shows the intercalates divided into three sets of size 4, so that each set partitions the set of 16 positions. Convince yourself that these sets can be regarded as the rows, columns and symbols of a second Latin square isomorphic to the first but with the positions labelled differently. The map taking the first set of labels to the second is an automorphism of the Latin square graph which is not an isotopism of the Latin square.

Another way to see the automorphism group in this case is to think of the graph as $L_2(4)$. It has $24 \cdot 24$ automorphisms permuting rows and columns of

A	B	C	D
B	A	D	C
C	D	A	B
D	C	B	A

1	2	3	4
5	6	7	8
9	10	11	12
13	14	15	16

1	2	5	6	1	3	9	11	1	4	13	16
3	4	7	8	2	4	10	12	2	3	14	15
9	10	13	14	5	7	13	15	5	8	9	12
11	12	15	16	6	8	14	16	6	7	10	11

Figure 6.10 Intercalates form a Latin square

the square array, and an automorphism transposing the array, giving the full automorphism group the order $24^2 \cdot 2 = 1152$ and structure $(S_4 \times S_4) \cdot C_2$.

In the case $n = 3$, the Latin square graph is the complete tripartite graph with sets of size 3; so we can permute the vertices within each part independently, and permute the three parts, giving an automorphism group $(S_3^3) : S_3$ with order $6^4 = 1296$.

We now give a solution in three of the six cases of the problem stated in Section 3.5, where we asked whether the edge set of the multigraph $K_{16}^{\{2\}}$ can be partitioned into i copies of the Shrikhande graph and $5 - i$ copies of $L_2(4)$, for $i = 0, 1, \ldots, 5$. Here we show that the answer is "yes" for $i = 0, 1, 2$.

We will call a vertex-disjoint union of four copies of K_4 a *factor* here. We note that two edge-disjoint factors on 16 vertices form the graph $L_2(4)$, while a Latin square graph consists of 3 edge-disjoint factors.

Let each of G and G' be either $L_2(4)$ or the Shrikhande graph. Their complements are Latin square graphs, as we have seen, and so are unions of three edge-disjoint factors, say F_1, F_2, F_3 (for G) and F_1', F_2', F_3' (for G'). Since any two of these factors form a grid $L_2(4)$, we can arrange the vertices so that $F_1 = F_1'$ and $F_2 = F_2'$. Then each of the three graphs $F_1' \cup F_2$, $F_2' \cup F_3'$ and $F_3 \cup F_1$ is isomorphic to $L_2(4)$, and these use all the edges of $K_{16}^{\{2\}} \setminus (G \cup G')$.

Since $\{G, G'\}$ contains 0, 1 or 2 copies of the Shrikhande graph, we have solved the problem in these cases.

Exercises

6.1 Use the tables for the field of order 4 on Fig. 6.6 in Section 6.4 to construct three mutually orthogonal Latin squares of order 4.

6.2 Show that the complement of a $PL(s, m)$ strongly regular graph is a $PL(m + 1 - s, m)$ graph.

6.3 Use the description of the Shrikhande graph given in Section 2.6 to identify it with the complement of the Latin square graph from the Cayley table of $\mathbb{Z}_4$.

6.4 This exercise constructs the partition of the edge set of K_{16} into three Petersen graphs. Let F be the finite field with 16 elements.

(a) Show that the multiplicative group of F is cyclic of order 15, and has the form $A \times B$, where A and B are cyclic groups of orders 5 and 3 respectively.

(b) Show that the sum of the elements in a proper non-empty subset of A is not zero. (Use the fact that the minimal polynomial of a non-trivial element of A is $x^4 + x^3 + x^2 + x + 1$.)

(c) Hence show that the Cayley graph $\mathrm{Cay}(F^+, A)$ is isomorphic to the Clebsch graph.

(d) Show that multiplication by elements of B maps the edge set of A to three pairwise disjoint sets of edges, each forming a graph isomorphic to the Clebsch graph.

The next three exercises concern the *Paley graphs*, an important family of graphs. Although they are strongly regular and have a lot of symmetry, they behave in other ways like random graphs, as we will see.

6.5 Let q be an odd prime power, and let F be the Galois field of order q.

(a) Show that half of the non-zero elements of F are squares and half are non-squares. [Hint: Multiplication by a non-square maps squares to non-squares and *vice versa*.]

(b) Show that -1 is a square if and only if q is congruent to 1 modulo 4.

Now let q be a prime power congruent to 1 mod 4, and F the field with q elements. The *Paley graph* $P(q)$ is defined as follows: the vertex set is F, and x and y are joined if and only if $y - x$ is a square.

6.6 (a) Show that $P(q)$ is an (undirected) graph.

(b) Show that $P(q)$ is self-complementary.

(c) Show that $P(q)$ is strongly regular, with parameters
$(q, (q-1)/2, (q-5)/4, (q-1)/4)$.

(d) Show that $P(5)$ is isomorphic to the 5-cycle graph, and $P(9)$ is isomorphic to $L_2(3)$.

6.7 Let q be an odd prime power which is a square, say $q = r^2$. Let G be the Paley graph $P(q)$.

(a) Show that G has clique number and independence number r.

(b) Show that G is both a pseudo-Latin square graph $PL_{(r+1)/2}(r)$ and a negative Latin square graph $NL_{(r-1)/2}(r)$.

(c) Show that in fact G is a Latin square graph.

Note: Finding the clique number of $P(q)$ when q is not a square is an open problem.

6.8 If q is a prime power congruent to 1 (mod 4), show that the edge set of K_{q+1} can be partitioned into two copies of the Paley graph $P(q)$.

6.9 Let q be a prime power congruent to 3 (mood 4), and let F be the field with q elements. Let H be the group

$$\{x \mapsto a^2 x + b : a, b \in F, a \neq 0\}.$$

Show that $|H| = q(q-1)/2$, and that the triangular graph $T(q)$ is a Cayley graph for the group H.

Remark Raymond Paley (1907–33) was a brilliant mathematician who worked in Cambridge and MIT, but was killed in an avalanche while skiing in Banff, Canada, at the age of 26. Figure 6.11 shows his tombstone in the cemetery in Banff. Paley never considered Paley graphs, but he was interested (among many other things) in the squares and non-squares in finite fields of

Figure 6.11 Tombstone of Paley in Banff

order congruent to 3 modulo 4, which are related to a class of Hadamard matrices now called Paley Hadamard matrices. See the paper [61] by Gareth Jones for the curious history of Paley graphs.

Bollobás and Thomason showed that a Paley graph $P(q)$ embeds all not-too-large graphs as induced subgraphs: specifically, if $q \geq (2^{r-2}(r-1) + 1)^2$, then $P(q)$ contains every r-vertex graph as an induced subgraph. This property holds with high probability in a random graph (with edges chosen independently with probability $1/2$); so Paley graphs are called *pseudo-random*.

7

The Shrikhande Graph on the Torus

Here is a very different picture of the Shrikhande graph from that we have seen before (Fig. 7.1).

The graph appears to have 25 vertices. But the purpose of the small single and double arrows is to tell us that the five vertices along the bottom of the parallelogram are to be identified with the five vertices along the top, while the five on the left are identified with the five on the right, keeping the same order in each case. (So for example the four vertices of the parallelogram become a single vertex after this identification.)

Before proceeding, we stop to identify the graph with the Shrikhande graph, using the description given in Section 2.6. Recall that the vertices are labelled with triples (x,y,z), where $x,y,z \in \mathbb{Z}$ and $x + y + z = 0$; and a step from a vertex to its neighbour is done by adding $(0,1,-1)$ or any of its permutations

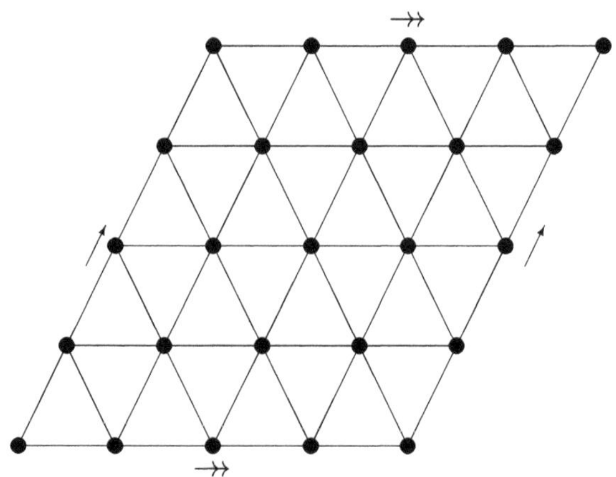

Figure 7.1 The Shrikhande graph on the torus

to the coordinates. Now label the vertex in the centre of the diagram in Fig. 7.1 $(0,0,0)$; and let the moves work as follows: $(1,0,-1)$ moves one step right; $(-1,0,1)$ moves one step left; $(0,1,-1)$ moves up and to the right; $(0,-1,1)$ moves down and to the left; $(-1,1,0)$ moves up and to the left; $(1,-1,0)$ moves down and to the left. Check that this uniquely and correctly labels the vertices in the diagram, allowing for wrapping around at the edges.

Warning: In our discussions of graphs such as $L_2(n)$ and $T(n)$, as shown in Fig. 3.5, we used the convention that two vertices on a line are to be regarded as joined. Here we have a different convention: as is more usual for a drawing of a graph, two vertices are joined only if there is a line segment joining them, with no intermediate vertices.

7.1 Orientable Surfaces

Now imagine that this figure is drawn on a rubber sheet. To make the identifications of top and bottom, we roll the sheet into a cylinder, distorting it so that the vertices on top and bottom coincide. Now we twist the cylinder so that the circles at the two ends also become identified.

The resulting surface is a *torus* or 'doughnut' (Fig. 7.2).

We can talk about graphs embedded in the torus in just the same way as we did for planar graphs when we were discussing Kuratowski's theorem (Theorem 2.5) in the introductory section. Just as there, if a graph is embeddable in the torus (or any surface), then so is any subdivision of it.

Although the Shrikhande graph cannot be drawn in the plane without edges crossing, it can be drawn on the torus.

Here, to test your geometric imagination, is a description of how to draw the Shrikhande graph on a physical torus.

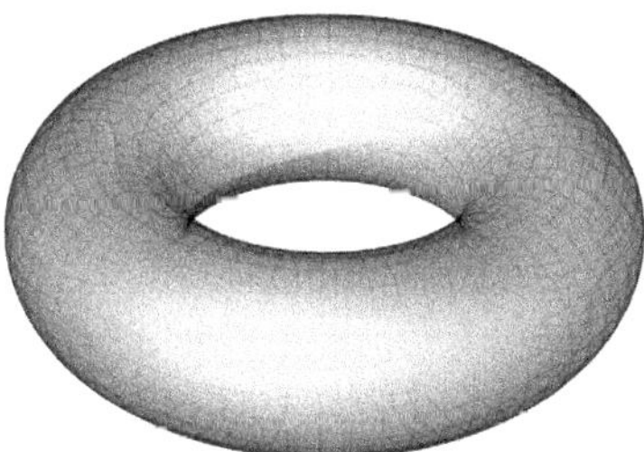

Figure 7.2 A torus (DemonDeLuxe (Dominique Toussaint), CC BY-SA 3.0, via Wikimedia Commons)

First, draw four circles passing through the central hole, by intersecting the torus with two perpendicular vertical planes through the central point. Now subdivide these circles each with four points, as follows: taking $0°$ to be the point of the circle furthest from the central point of the torus, put points at $0°$, $90°$, $180°$ and $270°$ on one circle, then at $45°$, $135°$, $225°$ and $315°$ on the next, and alternate for the other two. Finally, take a geodesic arc joining each vertex on one of these circles (at $d°$, say) to the vertices on the neighbouring circles at $(d \pm 45)°$. Check that the graph has diameter 2, degree 6; any two points have two common neighbours, and the induced subgraph on the neighbourhood of any vertex is a 6-cycle.

Exercise   (a) Draw a toroidal representation of K_5 and $K_{3,3}$.
(b) Draw the toroidal representation of the Shrikhande graph, as suggested.

Topology is popularly regarded as the study of properties unaffected by bending and stretching. The discussion here shows that we are in the realm of topology here. We are particularly interested in surfaces and give a rather informal treatment. We will use the term *homeomorphism* to mean a transformation which preserves the topological structure.

A surface is a connected space in which each point has a small neighbourhood which has the structure of a 2-dimensional disc. The surface of a sphere or a torus are examples of tori. These two surfaces are not homeomorphic. On the sphere, any closed curve can be shrunk to a point without leaving the surface; but on the torus, there are closed curves which cannot be shrunk to a point.

(We are describing here a surface without boundary. A disc is a surface with boundary, the circular rim forming the boundary; the points on the boundary have the property that a small neighbourhood is homeomorphic to a half-plane.)

A surface is said to be *orientable* if, whenever a circle enclosing a point is moved around the surface and returned to its starting position, its orientation remains the same; if this is not the case, the surface is *non-orientable*. The simplest example of a non-orientable surface is the *Möbius strip*, which can be produced by taking a strip of paper and giving one end a $180°$ twist, then glueing the two ends together. If a small circle is taken around the strip and returned to its initial position, its orientation is reversed.

There is a complete classification of surfaces. We deal with the orientable case here. We will not prove this theorem. A number of proofs exist; we refer to Seifert and Threlfall [92].

Theorem 7.1 *Let S be an orientable surface without boundary. Then there is a non-negative integer g such that S is homeomorphic to a sphere with g holes.*

The number *g* is called the *genus* of the surface. Thus, the sphere has genus 0, while the torus has genus 1.

An *embedding* of a graph *G* in a surface *S* is an embedding of the vertices and edges of *G* in *S* so that edges do not intersect except at their vertices. We note that embeddings on the sphere and on the plane are essentially the same thing, since if we have a graph embedded on the sphere, we can arrange so that the north pole of the sphere is not a vertex or on an edge of the graph; then we can puncture the sphere at the north pole and stretch it out so that it becomes a plane.

An embedding of a graph in a surface is *cellular* if its complement in the surface is homeomorphic to a number of disjoint discs; these are called the *faces* of the embedding.

Not all embeddings are cellular. Take, for example, a graph embedded in the sphere, and add a 'handle' to one face, converting the sphere into a torus. The face containing the handle is no longer a disc.

Embeddings of a graph in the plane are necessarily cellular; a graph is planar (according to our earlier definition) if it has an embedding in the plane (or equivalently on the sphere). We just have to recall that, for plane embeddings, there is an outside face which doesn't look like a disc until we reverse the aforementioned procedure to change the plane into a sphere. So a triangle drawn in the plane has two faces, an inside and an outside.

Theorem 7.2 *For any surface S, there is a number $\chi(S)$ such that, if a graph G with v vertices and e edges has a cellular embedding on S with F faces, then*

$$v - e + f = \chi(S).$$

Moreover, if S has genus g, then $\chi(S) = 2 - 2g$.

The number $\chi(S)$ is the *Euler characteristic* of *S*. For genus 0, this is *Euler's formula*.

Example: the Platonic solids Figure 7.3 shows the five famous Platonic solids. Regarded as graphs, they all have cellular embeddings on the sphere. (Draw the circumscribed sphere of the polyhedron, and project the edges

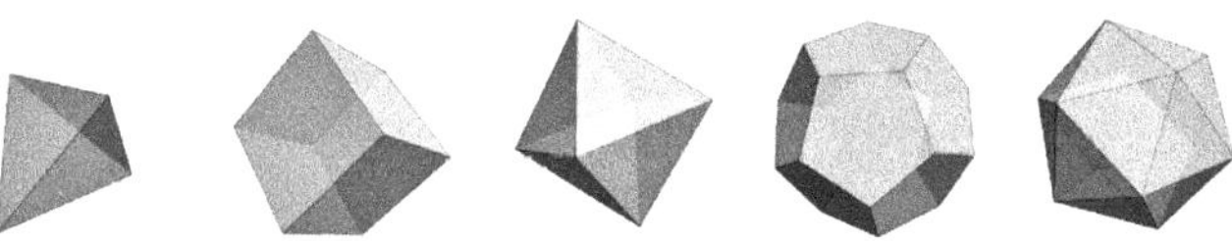

Figure 7.3 Regular polyhedra

outwards from the centre of the sphere onto its surface.) The numbers of vertices, edges and faces in each case are:

Tetrahedron: $v = 4$, $e = 6$, $f = 4$.
Cube: $v = 8$, $e = 12$, $f = 6$.
Octahedron: $v = 6$, $e = 12$, $f = 8$.
Dodecahedron: $v = 20$, $e = 30$, $f = 12$.
Icosahedron: $v = 12$, $e = 30$, $f = 20$.

Note that, in each case, $v - e + f = 2$; so the sphere has Euler characteristic 2 and genus 0.

Example: The Graph The drawing at the start of this chapter shows that the Shrikhande graph has a cellular embedding on the torus. The graph has 16 vertices and 48 edges. There are 6 faces surrounding each vertex, and since each face has 3 vertices, the number of faces is $16 \cdot 6/3 = 32$. Thus, the Euler characteristic of the torus is $16 - 48 + 32 = 0$, and the genus is 1.

Recall the definition of subdivision of a graph in Theorem 2.4, and Kuratowski's theorem: a graph is planar if and only if it does not contain a subdivision of K_5 or $K_{3,3}$ as a subgraph. It is known that there is a finite set $\mathscr{F}$ of graphs such that a graph is *toroidal*, or embeddable on the torus, if and only if it does not contain any subdivision of a graph in $\mathscr{F}$ as a subgraph. The set $\mathscr{F}$ is rather large, and it seems not to be explicitly known.

Definition The *genus* of a graph G is the smallest integer g such that G can be embedded in an orientable surface of genus g. We call this the *orientable genus* if it is necessary to distinguish it from the non-orientable genus (whose definition follows shortly).

Thus, a planar graph has genus 0, while the Shrikhande graph has genus 1.

The Dyck graph revisited From the representation of the Shrikhande graph on the torus, we obtain a visualization of the Dyck graph (see Chapter 5), showing that it is toroidal. The SG divides the torus into 32 triangular regions. Place a vertex in the centre of each triangle, and join two vertices in triangles sharing an edge; the result is the Dyck graph, with degree 3. We see that the Dyck graph is the dual of the Shrikhande graph on the torus, in the same sense that the dodecahedral graph is the dual of the icosahedral graph on the sphere.

We also see that the Dyck graph is bipartite. The triangles in the Shrikhande graph are of two types: "apex up" and "apex down"; two triangles sharing an edge have opposite types. (The consistency of this depends on the fact that the surface is orientable; otherwise the two types would be equivalent.) So the centres of the two types of triangles form a bipartition of the graph.

7.2 Non-orientable Surfaces

Non-orientable surfaces are a little more difficult to visualize.

We begin with a *Möbius band*, constructed as follows: take a long thin rectangle or streamer of paper. If you simply glue the ends together, you obtain a short wide cylinder. But if you give one end a 180° twist before glueing it to the other, you get a very different kind of surface. It has a boundary, which consists of a single closed curve; if a 2-dimensional creature on the surface were to walk around the strip and return, it would be on the other side of the strip, and so (assuming it is really just 2-dimensional) it would be the mirror image of what it was when it set out.

This is a non-orientable surface with a boundary. To produce such a surface without boundary, we have to glue the boundary to itself so that antipodal points are identified. This is somewhat difficult to imagine, so we give a different construction.

In Section 6.4.3, we met the projective plane over a finite field F: its points are the 1-dimensional subspaces of the 3-dimensional vector space over F, and lines are 2-dimensional subspaces. Here we are doing topology rather than geometry, so we just need the set of points. We simply replace the finite field F by the field $\mathbb{R}$ of real numbers. So we can visualise the points as lines through the origin in 3-dimensional Euclidean space. It is still not trivial to visualize a surface whose points are lines through the origin, so we take two more steps to obtain a convenient visualization:

(a) Take a sphere S with centre at the origin. Then each line through the origin meets S in two points, so we have a representation where each point of the projective plane is represented by two antipodal points on the sphere.

(b) Next we need a convenient way to choose one of each pair. Any antipodal pair either has one point in the northern hemisphere and one in the southern, or has both on the equator. For the first type, we use the points in the southern hemisphere as representatives. For the second type, we live with the ambiguity.

(c) Now if a small (2-dimensional) creature in the southern hemisphere crosses the equator, it appears in mirror image form on the opposite point of the equator.

So our surface consists of the southern hemisphere with a convention about crossing the boundary.

A note about reflection. When you look in a mirror, your image is reflected front-to-back. But because humans are not front-to-back symmetric but are approximately left-to-right symmetric, you interpret what you see as a left-to-right reflection combined with a half-turn. (This is the answer to the common question 'Why does a mirror reflect left-to-right but not top-to-bottom?') In the same way, the small creature crossing the equator heading north in our model of the real projective plane finds itself translated to the opposite point on the equator, reflected, and heading south.

Since we are doing topology, we can regard the southern hemisphere as a rubber sheet, and pull it up so that the equator becomes a small circle at the north pole. Now the point set is almost the whole sphere, but the rule for crossing the circle remains the same. Someone crossing the circle at one point reappears reflected at the opposite point. We call this circle a *crosscap*. Note that the interior of the small circle no longer exists as an accessible part of the surface.

Now the real projective plane is a sphere with a single crosscap (which in our construction is at the north pole, but could be anywhere).

More generally, we can take a sphere, and replace any positive number k of pairwise disjoint discs by crosscaps; we obtain a non-orientable surface.

With this definition, we can state the characterization of non-orientable surfaces, and the analogue of Euler's formula for graphs embedded in such a surface.

Theorem 7.3 *A non-orientable compact surface without boundary is homeomorphic to a sphere with k crosscaps, for some k > 0.*

The number k is the *non-orientable genus* of the surface.

Definition   The *non-orientable genus* of a graph G is the smallest integer k such that G can be embedded in an non-orientable surface of genus k.

Theorem 7.4 (Euler's formula) *If a connected graph with v vertices and e edges has a cellular embedding in a sphere with k crosscaps, and the embedding has f faces, then*

$$v - e + f = 2 - k.$$

Example   Here is an example to show that the class of graphs embeddable in the real projective plane is larger than the class of graphs embeddable in spheres.

Let G be a graph embeddable in the sphere with a unique pair of edges which cross at a point p. Put a small crosscap at p. Then one can follow each edge without meeting the other, so we have a proper embedding.

The number $2 - k$ is the *Euler characteristic* of the non-orientable surface of genus k.

Definition The *non-orientable genus* of a graph G is the smallest integer g such that G can be embedded in an orientable surface of genus g.

Thus, a planar graph has genus 0, while the Shrikhande graph has genus 1.

7.3 Regular Maps

The regular polyhedra, as graphs, can be drawn on the sphere. (Take the circumscribed sphere of the polyhedron, passing through all the vertices; then project the edges outwards onto the surface of the sphere.) The resulting embedding has the properties that any face has the same number of edges, and any vertex has the same degree.

To generalize this property, we first strengthen it. Each regular polyhedron has a large group, which is transitive on vertices, edges, and faces. More is true. A *flag* is an incident vertex-edge-face triple. (In Fig. 7.4, the vertex and edge are doubled, the face shaded.)

An automorphism of a map is required to map vertices to vertices, edges to edges, and faces to faces; this is a more stringent condition than just being a graph automorphism.

Now the automorphism group of a regular polyhedron is *flag-transitive*, that is, transitive on flags. But only the identity can fix a flag: for the other end of the edge of the fixed flag and all the vertices on the face are fixed, and working our way outwards, using the connectedness of a graph, we see that an automorphism fixing a flag is the identity. It follows from the Orbit–Stabilizer Theorem that the order of the group is equal to the number of flags.

A cellular embedding of a graph on a surface with a flag-transitive automorphism is called a *regular map*.

Suppose that a regular polyhedron has vertex degree r and face size s. The five regular polyhedra realize the values $(r, s) = (3,3), (3,4), (4,3), (3,5)$ and $(5,3)$.

The next smallest values are $(6,3), (4,4)$ and $(3,6)$. A little thought shows that these correspond to the plane tessellations by triangles, squares and hexagons respectively. See Fig. 7.5.

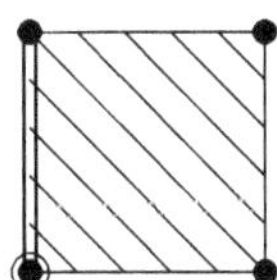

Figure 7.4 A flag

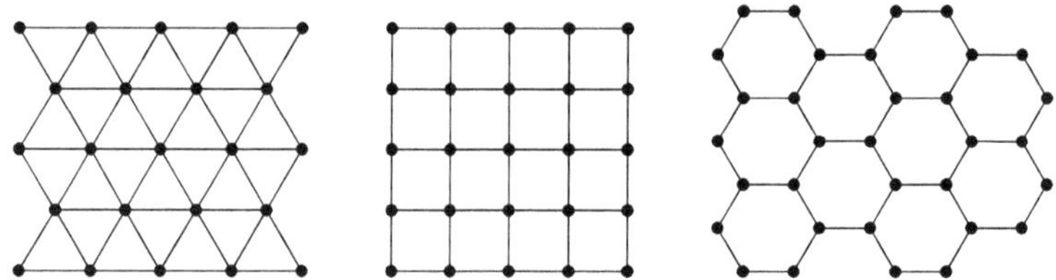

Figure 7.5 Regular tessellations of the plane

But these are not the only ones.

Theorem 7.5 *The Shrikhande graph is a regular map; it has valency 6 and the faces are triangles.*

For there are 16 vertices, each lying in 6 edges, and each edge in 2 flags, so altogether 192 flags; but we have seen that the Shrikhande graph has 192 automorphisms, and it is not hard to see that all are map automorphisms. (Every face is a triangle, so the faces are preserved by graph automorphisms.)

The torus can be formed by 'rolling up' the plane, identifying points which can be reached by combinations of two fixed translations. If we take the translations to be through four times the side length of a triangle in two directions at 60° in the tessellation of the plane by triangles, we obtain the Shrikhande graph.

For more information about graphs drawn on surfaces, we recommend [79].

Exercises

7.1 Suppose that a graph can be drawn on the sphere with only one pair of edges crossing. Show that it can be drawn on the torus. [Hint: hire a firm of civil engineers to build a bridge or tunnel; this converts the sphere to a torus.]

  Is the converse true?

7.2 Show that K_5 and $K_{3,3}$ cannot be drawn in the plane. (Hint: Use Euler's formula. What are the possible sizes of faces?)

7.3 Suppose that a regular polyhedron has the properties that each vertex lies on r edges and each face has s edges. Show using Euler's formula that

$$\frac{1}{2} + \frac{1}{r} + \frac{1}{s} > 1.$$

Deduce that the only regular polyhedra are the five known to the Greeks: the tetrahedron, cube, octahedron, dodecahedron and icosahedron.

8

Root Systems

Root systems are beautiful geometric objects in Euclidean space, which crop up in many parts of mathematics, including Lie algebras, singularity theory, mathematical physics and graph theory.

In this chapter, we will use graph theory to discuss the famous *ADE classification* of root systems in which all roots have the same length. This will then be used to determine the graphs whose adjacency matrix has least eigenvalue -2 or greater in Chapter 9, where we will see a connection between the Shrikhande graph and exceptional root systems.

8.1 The Definition

We work in the Euclidean space $V = \mathbb{R}^d$, with the standard inner product.

Given a non-zero vector $u \in V$, there is a unique hyperplane H_u through the origin which is perpendicular to u. We define the *reflection r_u* in this hyperplane to be the linear map which fixes every vector in H_u and maps u to $-u$.

The formula for this reflection is

$$r_u(x) = x - \frac{2(x.u)}{u.u}\, u.$$

This map is clearly linear and does map u to $-u$ and fixes every x satisfying $x.u = 0$.

A *root system* is a set S of non-zero vectors of $\mathbb{R}^d$ satisfying the four conditions

- $\langle S \rangle = V$ (S spans V);
- if $u, \lambda u \in S$, then $\lambda = \pm 1$;
- if $u, v \in S$, then $2(v.u)/(u.u)$ is an integer;
- for all $u \in S$, the reflection r_u maps S to itself.

We remark that the first condition is not crucial, since we could replace V by the subspace spanned by S to ensure that it holds. Also, the fourth condition implies the converse of the second (that is, if $u \in S$, then $r_u(u) = -u \in S$). Also, the third condition shows that $r_u(v)$ is an integer combination of u and v, for all $u, v \in S$. (This is called the *crystallographic condition*.)

Suppose that S_1 and S_2 are root systems in spaces V_1 and V_2. Then clearly $S_1 \cup S_2$ is a root system in the orthogonal direct sum of V_1 and V_2. Such a root system is called *decomposable*; if there is no orthogonal direct sum decomposition with S the union of its intersections with the two summands, it is called *indecomposable*. Clearly, to classify all root systems, it suffices to classify the indecomposable ones, and take direct sums of them.

It turns out that the indecomposable root systems fall into four infinite families apart from five sporadic examples. We are not going to prove this, but will deal with the case where all the roots have the same length. First, here are the examples. We assume that $e_1, e_2, \ldots, e_n$ is an orthonormal basis for $\mathbb{R}^n$.

- $A_n = \{e_i - e_j : 1 \le i,j \le n+1, i \ne j\}$.
- $D_n = \{\pm e_i \pm e_j : 1 \le i < j \le n\}$.
- $E_8 = D_8 \cup \{\frac{1}{2} \sum_{i=1}^{8} \epsilon_i e_i : \epsilon_i = \pm 1, \prod_{i=1}^{8} \epsilon_i = +1\}$.
- $E_7 = \{u \in E_8 : (e_1 - e_2).u = 0\}$.
- $E_6 = \{u \in E_8 : (e_1 - e_2).u = (e_2 - e_3).u = 0\}$.

It can be verified that each of these is a root system; the subscript gives the dimension. (Note that the vectors of A_n do not span the whole of $\mathbb{R}^{n+1}$, since each of them is orthogonal to the vector $e_1 + \cdots + e_{n+1}$; they span the n-dimensional space perpendicular to this vector.) See Exercise b.

With a little more effort we can see that D_2 is decomposable (it consists of two vectors in each of two perpendicular directions), and D_3 is isomorphic to A_3; hence we consider D_n only for $n \ge 4$.

Root systems are beautiful symmetrical objects: Mark Ronan, in his book on the history of the finite simple group classification, refers to them as *multidimensional crystals*. Figure 8.1 shows A_2 and A_3.

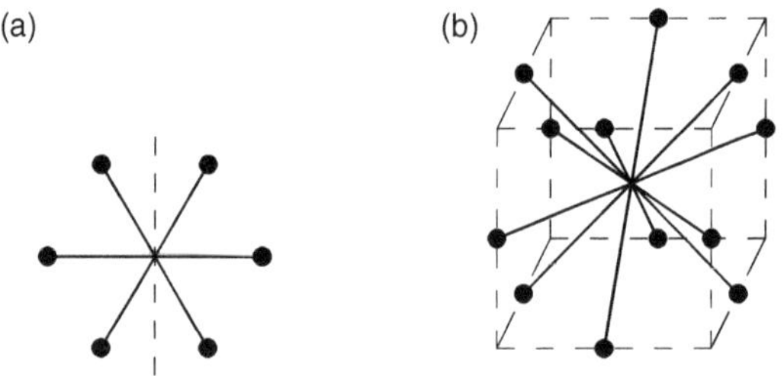

Figure 8.1 The root systems A_2 and A_3

Theorem 8.1 *An indecomposable root system in which all the roots have the same length is isomorphic to A_n ($n \geq 1$), D_n ($n \geq 4$), E_6, E_7 or E_8.*

We will prove this theorem after a small diversion. Using this result, it is not too hard to extend the theorem to deal with arbitrary root systems; this is outlined in Exercise 3.6.

8.2 A Graph-Theoretic Result

Now we recall Theorem 3.4. We say that a graph G has the *strong triangle property* if

> Every edge $\{u,v\}$ of G is contained in a triangle $\{u,v,w\}$ with the property that, for any vertex $x \notin \{u,v,w\}$, x is joined to exactly one of u,v,w.

Theorem 8.2 *A graph with the strong triangle property is one of the following:*

(a) a null graph;
(b) a Friendship Theorem graph; see Fig. 3.1;
(c) one of three special graphs on 9, 15 and 27 vertices.

Remark The 9-vertex graph is $L_2(3)$, and the 15-vertex graph is the complement of $T(6)$. It is straightforward to check that the graphs in (a) and (b) do indeed satisfy the strong triangle property. For (c), the 9-vertex graph is $L_2(3)$, the 15-vertex graph is the complement of $T(6)$ and the 27-vertex graph is the Schläfli graph, which was discussed (and an explicit construction given) in Section 3.2.3.

Proof Let G be such a graph. If G has no edges, then (a) holds, so we may assume there is at least one edge. Now we follow the proof of the Friendship Theorem, by showing that either (b) holds or the graph is regular.

First we claim that, if u and v are not joined, then they have the same valency. For, given any vertex u, the subgraph on the neighbours of u is a Friendship Theorem graph, consisting of (say) m triangles joined at a vertex, and so u has valency $2m$. Since v is not joined to u, it is joined to one vertex in each of these triangles; so the neighbourhood of v consists of at least m triangles, and v has valency at least $2m$. Reversing the roles of u and v gives the claim.

Suppose that the graph is not regular, and let a and b be vertices with different valencies. There is a third vertex c joined to both, as in the strong triangle property. Now c has different valency from at least one of a and b, say a. Any further vertex is joined to exactly one of a,b,c, and also to at least one of a and b and at least one of a and c; so it is joined to a, and we have a Friendship Theorem graph.

In the case when the graph is regular, we have that any two adjacent vertices have one common neighbour, while two non-adjacent vertices have m common neighbours, where $2m$ is the valency of the graph. So it is strongly regular, with parameters $(n, 2m, 1, m)$; and calculation shows that $n = 6m - 3$.

Our analysis of strongly regular graphs in Chapter 1 shows that the eigenvalues are $2m$, 1 and $-m$. If their multiplicities are $1, f, g$, we get

$$1 + f + g = 6m - 3,$$

$$2m + f - mg = 0,$$

so $(m + 1)g = 8m - 4$. Thus $m + 1$ divides 12, that is, $m = 2, 3, 5$ or 11.

Further analysis shows that $m = 11$ is impossible while the graphs in the other three cases are unique: see Exercise 3.2. (The argument there comes up with the values 2, 3, 5 without using the strongly regular graph conditions.)

8.3 Proof of Theorem 8.1

Suppose we have an indecomposable root system in $\mathbb{R}^d$ with all roots of the same length. Without loss of generality, we take this length to be $\sqrt{2}$; so $u.u = 2$ for every root u, and $u.v \in \{2, 1, 0, -1, -2\}$ for any roots u and v. This means that any two roots are at an angle $0°$, $60°$, $90°$, $120°$ or $180°$. In other words, the lines spanned by the roots make angles $90°$ or $60°$ with each other.

There exist two roots u, v with $u.v = -1$ (in other words, an angle $120°$). Then $w = -u - v$ is also a root (it is $-r_u(v)$). We call six roots of the form $\pm u$, $\pm v$ and $\pm w$ a *star*. (They form a root system of type A_2 in the plane they span.)

Let S be a set of lines through the origin in Euclidean space $\mathbb{R}^d$, any two making an angle $90°$ or $60°$. We say that S is *star-closed* if, whenever two lines $L_1, L_2 \in S$ are at angle $60°$, the third line in their plane making an angle $60°$ with both is also in S. (See Fig. 8.2.)

Proposition 8.3 *Fix a positive number l. Then the vectors of length l in both directions along the lines of a set S form a root system if and only if the lines in S make angles $90°$ or $60°$ with each other and the set S is star-closed.*

This is hopefully obvious from the preceding remarks. We will classify the root systems by classifying the line systems with these properties instead.

If $\{\langle u \rangle, \langle v \rangle, \langle w \rangle\}$ is a star with $u + v + w = 0$, then any further root is orthogonal to one or all of u, v, w. For $x.u, x.v, x.w \in \{-1, 0, 1\}$ and $x.u + x.v + x.w = 0$.

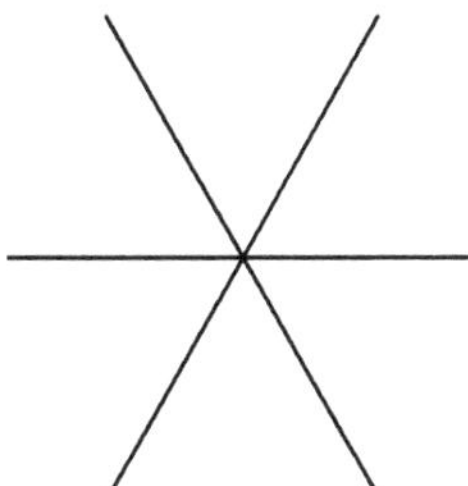

Figure 8.2 A star

Let A_u, A_v, A_w, B be the sets of lines spanned by roots which are orthogonal to just u, just v, just w, or all three, and choose spanning roots on these lines so that, for example, if $\langle x \rangle \in A_u$, then $x.v = +1$ and $x.w = -1$. Now form a graph G with vertex set A_u, in which two vertices are adjacent if the corresponding lines are perpendicular.

First we claim that any two spanning vectors of lines in A_u have non-negative inner product. For if x,y are two such and $x.y = -1$, then $\langle x + y \rangle \in A_u$, and $(x + y).v = 2$, which forces $x + y = v$, a contradiction.

Next we claim that the graph G has the strong triangle property. For suppose that $\langle x \rangle$ and $\langle y \rangle$ are adjacent, so that $x.y = 0$. Then $v - x$ and $w + y$ are roots, and $(v - x).(w + y) = 1$; so $v - x - w - y = z$ is a root. Checking inner products, we find that $\langle z \rangle \in A_x$. Now $x + y + z = v - w$, and so

$$(t.x) + (t.y) + (t.z) = 2$$

for all $\langle t \rangle \in A_u$. This means that exactly two of these three inner products are 1, so exactly one of the lines $\langle x \rangle$, $\langle y \rangle$ and $\langle z \rangle$ is perpendicular to $\langle t \rangle$ (and so joined to it in G).

Finally we claim that G determines the root system uniquely. For the graph structure of G determines the inner products of vectors spanning the lines in

$$\{\langle u \rangle, \langle v \rangle, \langle w \rangle\} \cup A_u;$$

and any other root can be obtained from these by closing under reflection.

Now it is readily checked that, for the root systems A_n, D_n, E_6, E_7 and E_8, the structure of the graph G is null, a Friendship graph, or one of the three exceptional graphs in Theorem 3.4. This completes the proof.

For example, take the root system A_n, and let $u = e_1 - e_2$, $v = e_2 - e_3$ and $w = e_3 - e_1$. Then A_u consists of the lines spanned by vectors $e_2 - e_i$ for $i > 3$, and no two of these are orthogonal, so G is the null graph.

8.4 Another Approach to Root Systems

Root systems first arose in the classification of simple Lie algebras over the complex numbers. The classification theorem was proved by a different method from the one we described; but there is a very nice connection.

The classification depends on the following. A *fundamental basis* for a root system R is a subset B of R with the two properties

(a) every root in R can be written uniquely as an integer linear combination of the roots in B, with either all coefficients ≥ 0, or all coefficients ≤ 0;
(b) any two roots in B have inner product ≤ 0.

In fact the second condition follows from the first. For, if v, w are roots of the same length making an angle of $60°$, then $v - w$ is a root, and its coefficients include both positive and negative.

Theorem 8.4 *Every root system has a fundamental basis.*

We don't prove this; instead we describe how we can proceed from there.

Let R be a root system with all roots of the same length (which, by scaling, we can take to be $\sqrt{2}$) and B a fundamental basis. Then any two roots in B have inner product 0 or -1, that is, angle $90°$ or $120°$. Then the matrix of inner products of the vectors in B has diagonal entries 2 and off-diagonal entries 0 or -1; that is, it has the form $2I - A$, where A is the adjacency matrix of a graph. Now this matrix is positive semi-definite, so all eigenvalues of A are 2 or smaller. Moreover, we can assume that the graph is connected, since otherwise it is a disjoint union of connected components, all of which have all eigenvalues 2 or smaller.

In fact, our conclusion is a bit stronger. Since B is a basis, the inner product matrix $2I - A$ is non-singular, hence positive definite (no eigenvalue zero); so A has all eigenvalues strictly less than 2.

We will also use the following result, whose proof is deferred to the end:

Theorem 8.5 *The largest eigenvalue of an induced subgraph of a graph is bounded above by the largest eigenvalue of the whole graph.*

Our main result is as follows.

Theorem 8.6 *A connected graph whose eigenvalues are all smaller than 2 is one of those shown in Fig. 8.3.*

These graphs are famously known as the *Coxeter–Dynkin diagrams*, or *Dynkin diagrams* of type ADE, or the *ADE diagrams*. They occur throughout

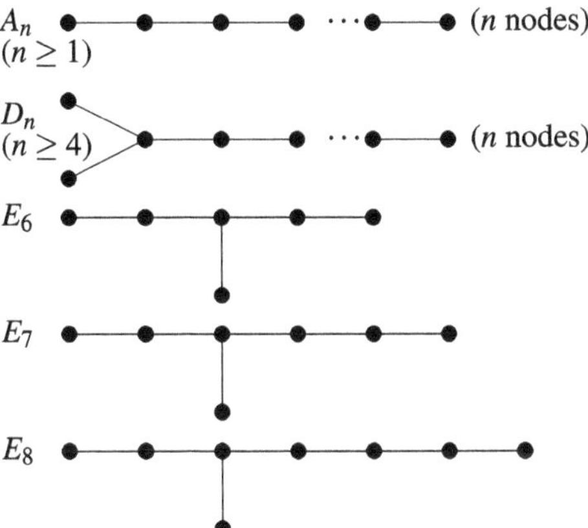

Figure 8.3 The ADE diagrams

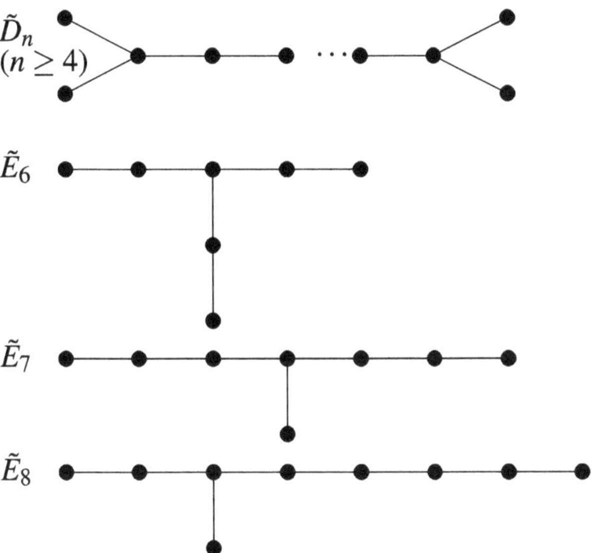

Figure 8.4 Extended ADE diagrams

mathematics, from instants in mathematical physics through cluster algebras and singularities of smooth functions to regular polytopes and finite reflection groups.

In order to prove this we need to consider another class of graphs, the *extended ADE diagrams*, denoted by putting a tilde over the symbol for the ordinary diagram. The extended A_n, thus denoted $\tilde{A}_n$, is a cycle with $n + 1$ vertices. The others are shown in Fig. 8.4.

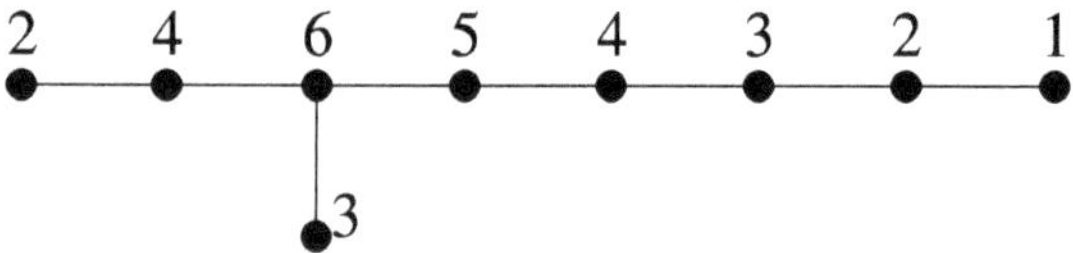

Figure 8.5 An eigenvector for $\tilde{E}_8$

Each extended diagram contains the ordinary diagram as an induced subgraph but has one extra vertex.

Lemma 8.7 *Each extended ADE diagram has eigenvalue* 2.

An eigenvector with eigenvalue 2 is an assignment of a number to each vertex (not all the numbers being zero) such that the sum of the labels on the neighbours of any vertex is twice the label of the vertex itself. For $\tilde{A}_n$, a cycle, we label each vertex with 1. The others are an exercise for the reader. Figure 8.5 shows the labelling for $\tilde{E}_8$.

So let Γ be a connected graph whose eigenvalues are all strictly less than 2. Then Γ can't contain $\tilde{A}_n$, a cycle; so it is a tree. Also Γ can't contain $\tilde{D}_n$ (for $n \geq 5$), so it has at most one vertex of valency greater than 2; and it can't contain $\tilde{D}_4$, so there is no vertex of valency 4. If every vertex has valency 2 except the leaves, then Γ is a path, that is, A_n. Otherwise it has a point of valency 3, from which arms of length $p - 1, q - 1, r - 1$ radiate (so p, q, r are the numbers of vertices on the arms, including the branch point). Since Γ can't contain $\tilde{E}_6$, $\tilde{E}_7$, $\tilde{E}_8$, we see that the possibilities for (p, q, r) are $(2, 2, n)$ (giving D_n) or $(2, 3, r)$ with $r = 3, 4, 5$ (giving E_6, E_7, E_8 respectively).

This proves the theorem.

We have on the way proved another rather surprising result.

Theorem 8.8 *If the eigenvalues of a connected graph Γ are not all strictly smaller than* 2, *then Γ has an induced subgraph which has an eigenvalue* 2.

This is because Γ must contain an extended ADE diagram as an induced subgraph. The obvious generalization, replacing 2 by other values, is not true.

If you have met the finite rotation groups in 3-dimensional Euclidean space (the cyclic and dihedral groups and the rotation groups of the tetrahedron, octahedron and icosahedron), you might notice some interesting coincidences here. Indeed the dihedral and polyhedral groups are defined by

$$\langle x, y, z : x^p = y^q = z^r = xyz = 1 \rangle,$$

with p, q, r, as in the aforementioned proof. If we use instead the values for the extended ADE diagrams, we get the rotation groups of regular tessellations of the plane.

There is more magic here, but we will stop now and turn to the proof of Theorem 8.5, which involves a useful argument from linear algebra.

8.5 Proof of Theorem 8.5

Lemma 8.9 *The largest eigenvalue of a real symmetric matrix A is equal to the maximum value of $v^\top Av/v^\top v$ over all non-zero vectors v.*

Proof Write v in a basis of eigenvectors for A:

$$v = \sum \alpha_i v_i \text{ where } Av_i = \lambda_i v_i.$$

Then

$$v^\top Av = \sum \lambda_i \alpha_i^2 |v_i|^2 \le \lambda_1 \sum \alpha_i^2 |v_i|^2 = \lambda_1 v^\top v,$$

where λ_1 is the largest eigenvalue of A. But if w is an eigenvector for the eigenvalue λ_1, then $Aw = \lambda_1 w$, and so

$$w^\top Aw = \lambda_1 w^\top w.$$

So the largest value of the ratio is λ_1.

Now suppose that Δ is an induced subgraph of Γ. Let A be the adjacency matrix of Γ, and λ_1 its largest eigenvalue. Let v be any vector whose coordinates are zero at all positions except those in the induced subgraph Δ. Then $v^\top Av \le \lambda_1 v^\top v$. But we can choose w to be an eigenvector of the adjacency matrix of Δ corresponding to the largest eigenvalue μ_1, extended by zeros everywhere else; then $Aw = \mu_1 w$, and so $w^\top Aw = \mu_1 w^\top w$. So $\mu_1 \le \lambda_1$ and the theorem is proved.

For later use, we give explicit realizations of the root systems. In each case, the vectors $e_1, e_2, \ldots$ form an orthonormal basis for the real vector space.

A_n: $\{e_i - e_j : 1 \le i,j \le n + 1, i \ne j\}$ (these vectors lie in the subspace of the $(n + 1)$-dimensional space $\langle e_1, e_2, \ldots, e_{n+1}\rangle$ consisting of vectors $\sum x_i e_i$ subject to the condition $\sum x_i = 0$: these lie in a codimension-1 subspace.)

D_n: $\{\pm e_i \pm e_j : 1 \le i,j \le n, i \ne j\}$.

E_8: $D_8 \cup \{\frac{1}{2} \sum x_i e_i : x_i = \pm 1, \prod x_i = 1\}$.

E_7: the subset of E_8 orthogonal to a fixed root.

E_6: the subset of E_8 orthogonal to three fixed roots forming a star.

For E_7 and E_6, the choice of the root or star does not affect the result, since, for example, any root can be transformed into any other by a suitable product of reflections.

From these explicit representations, we can read off a couple of inclusions:

(a) $A_n \subseteq D_{n+1}$. Indeed, A_n is the intersection of D_{n+1} with the hyperplane

$$\{x_1 e_1 + \cdots + x_{n+1} e_{n+1} : x_1 + \cdots + x_{n+1} = 0\}.$$

(b) $E_6 \subseteq E_7 \subseteq E_8$.
(c) $D_8 \subseteq E_8$.

The last two are clear from the representations given.

It is also true that $A_8 \subseteq E_8$. This can be seen by using a different representation of E_8:

$$E_8 = A_8 \cup \{f - e_i - e_j - e_k : \{i, j, k\} \subset \{1, \ldots, 9\}\},$$

where $f = \frac{1}{3}(e_1 + e_2 + \cdots + e_9)$. Note that the added vectors also have coordinate sum 0, so lie in the hyperplane spanned by the vectors of A_8.

Exercises

8.1 Using the description of the Schläfli graph given in Chapter 3, show that the Schläfli graph is strongly regular, with parameters $(27, 10, 1, 5)$, and that it has the strong triangle property.

8.2 Show that the examples given here are root systems. Show also that A_n spans a real vector space whose dimension is n, not $n + 1$.

8.3 Show that the two representations of E_8 given here are isomorphic.

9

Graphs with Least Eigenvalue -2

In this chapter, we use the classification of the root systems with all roots of the same length to determine the graphs G whose adjacency matrix $A(G)$ has least eigenvalue -2 or larger. This classification includes the Shrikhande graph and gives us another, quite different, proof of Shrikhande's Theorem.

Note that if G is a graph with at least one edge, then the sum of the eigenvalues of $A(G)$ is zero (the trace of $A(G)$), and hence G has both positive and negative eigenvalues. So, for non-null graphs, the smallest eigenvalue is negative.

9.1 Preliminaries

Theorem 9.1 *Let $A = (a_{ij})$ be a positive semidefinite real symmetric $n \times n$ matrix. Then there are vectors $v_1, \ldots, v_n \in \mathbb{R}^d$ with $v_i.v_j = a_{ij}$ for all i and j, where d is the rank of A.*

Proof From the theory of real quadratic forms we know that, if A is a real symmetric matrix, then there exists an invertible matrix P such that

$$PAP^\top = \begin{pmatrix} I_r & O & O \\ O & -I_s & O \\ O & O & O \end{pmatrix},$$

where $r + s$ and $r - s$ are the rank and signature of A. Since our matrix A is positive semidefinite, we have $s = 0$, and so

$$PAP^\top = \begin{pmatrix} I_d & O \\ O & O \end{pmatrix}.$$

Put $Q = P^{-1}$. Then

$$A = Q \begin{pmatrix} I_d & O \\ O & O \end{pmatrix} Q^\top = Q_1 Q_1^\top,$$

117

where Q_1 is the matrix consisting of the first d columns of Q. Now if v_i denotes the ith row of Q_1, the (i,j) entry of $A = Q_1 Q_1^\top$ is equal to $v_i.v_j$, as required.

Here is an application, to point us in the direction we will go.

Proposition 9.2 *Let G be a connected graph with smallest eigenvalue −1 (or greater). Then G is a complete graph.*

Proof Let $A(G)$ be the adjacency matrix of G. By assumption, $A(G) + I$ is positive semidefinite, so there are vectors $v_1, \ldots, v_n$ with $v_i.v_j$ equal to the (i,j) entry. Thus we have

- $v_i.v_i = 1$, so v_i is a unit vector.
- if the ith and jth vertices are adjacent, then $v_i.v_j = 1$. But then $(v_i - v_j).(v_i - v_j) = 0$, and so $v_i = v_j$.

Since the graph is connected, all the vectors v_i are equal. So $v_i.v_j = 1$ for all pairs (i,j); this means that G is a complete graph.

Indeed, the complete graph K_n has eigenvalues $n - 1$ (with multiplicity 1) and -1 (with multiplicity $n - 1$); see the Exercises.

9.2 Generalized Line Graphs

In this section we define generalized line graphs. The definition is due to Alan Hoffman, who proved the first version of the theorem given in the next section.

Our strategy is to represent a graph (if possible) by a set of vectors in Euclidean space, where $v_i.v_i = 2$ for all i, and

$$v_i.v_j = \begin{cases} 1 & \text{if vertices } i \text{ and } j \text{ are joined,} \\ 0 & \text{otherwise.} \end{cases}$$

Proposition 9.3 *A graph has a Euclidean representation in the above sense if and only if its adjacency matrix has least eigenvalue −2 or greater.*

Proof Suppose the least eigenvalue of $A(G)$ has least eigenvalue −2 or greater. Then $2I + A(G)$ is positive semidefinite, and so represents the inner product of a set of vectors forming the required Euclidean representation. The other direction works in the same way.

So which graphs can be represented?

First we observe that any line graph has such a representation. Let G be a graph on n vertices $1, 2, \ldots, n$. Take $e_1, e_2, \ldots, e_n$ to be an orthonormal basis for $\mathbb{R}^n$. Then, for each edge ij in the graph G, we take the vector $e_i + e_j$. It is clear that each vector has inner product 2 with itself, 1 with the vector $e_i + e_k$ representing an edge sharing a vertex with ij, and 0 otherwise; so our conditions are satisfied.

Note that the representing vectors all lie in the root system D_n.

Hoffman observed another class of graphs which are representable. The *cocktail party graph* $CP(n)$ has $2n$ vertices, numbered $1, 2, \ldots, 2n$; every pair of vertices is joined by an edge except for the pairs $(2i-1), 2i$ for $i = 1, 2, \ldots, n$. (The name reflects a cocktail party attended by n couples; each person talks to everyone at the party except for his/her partner.)

The cocktail party graph can be represented in $\mathbb{R}^{n+1}$, using the standard basis $e_0, e_1, \ldots, e_n$, in the following way: $2i - 1 \mapsto e_0 + e_i$, $2i \mapsto e_0 - e_i$. Again we have a representation in a root system, this time D_{n+1}.

Hoffman combined these into the notion of a *generalized line graph*, defined as follows. Given a graph G with vertex set $1, 2, \ldots, n$, and n non-negative integers $a_1, a_2, \ldots, a_n$, we define the graph $L(G; a_1, \ldots, a_n)$ as follows: take the disjoint union of the line graph of G with cocktail parties $CP(a_1), \ldots, CP(a_n)$, with some extra edges: a vertex of $L(G)$ corresponding to the edge ij in the graph G is joined to all vertices of the cocktail parties $CP(a_i)$ and $CP(a_j)$. Figure 9.1 shows a graph G and the generalized line graph $L(G; 2, 1, 0, 3)$.

A generalized line graph also has a Euclidean representation in the root system D_m, where $m = n + \sum_{i=1}^{n} a_i$, as follows. We label the basis $e_{i,k}$ where i runs from 1 to n, and for each i, k runs from 0 to a_i. Now for each edge ij of G, we choose the vector $e_{i,0} + e_{j,0}$ (these vectors represent the line graph of G). Then we add all edges $e_{i,0} \pm e_{i,k}$ where k runs from 1 to a_i; these represent the cocktail party $CP(a_i)$, and it is easy to see that the extra edges are handled correctly as well.

Now we can state the main theorem.

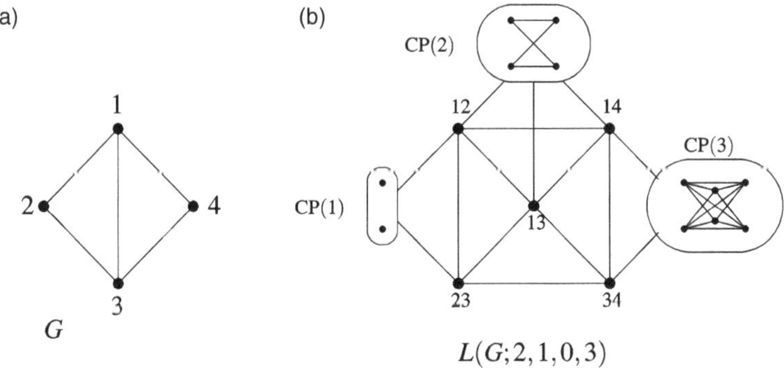

Figure 9.1 A graph and its generalized line graph

Theorem 9.4 *Let G be a connected graph with least eigenvalue −2. Then either*

(a) G is a generalized line graph; or
(b) G is represented by a subset of the root system E_8.

Remark In Hoffman's earlier version of the theorem, condition (b) simply required that the number of vertices of G be bounded above by a constant. In fact, it can be shown easily that a graph represented by a subset of E_8 has at most 36 vertices, and has valency at most 28; these bounds are best possible.

The proof depends on the following lemma.

Lemma 9.5 *Let S be a set of vectors in Euclidean space such that*

(a) $v.v = 2$ for all $v \in S$;
(b) $v.w \in \{-1, 0, 1\}$ for all $v, w \in S$ with $w \neq \pm v$.

Then S is a subset of a root system.

Proof We can adjoin to S the negatives of its vectors. Now suppose that $v, w \in S$ with $v.w = -1$. If $v + w \notin S$, we consider inner products:

- $(v + w).(v + w) = 2 - 2 + 2 = 2$.
- Take any vector $x \in S$. Then $x.(v + w) = x.v + x.w$, and each of $x.v$ and $x.w$ is in $\{-1, 0, 1\}$. If $x.v = x.w = 1$, then $x.(v + w) = 2$, and so

$$(x - v - w).(x - v - w) = 2 - 2 - 2 + 2 = 0,$$

so $x = v + w$, contrary to the assumption that $v + w \notin S$. A similar argument holds if $x.v = x.w = -1$. We conclude that we can add $v + w$ (and its negative) to S while preserving the hypotheses.

When this process terminates, we have a root system.

Proof of Theorem 9.4 To prove the theorem, take a graph with least eigenvalue −2 or greater. We know that it has a Euclidean representation, by a subset of a root system, and so we only need to examine the root systems. While giving the proof of Theorem 8.5, we observed that A_n is contained in D_{n+1} and $E_6 \subseteq E_7 \subseteq E_8$; so we can assume that the root system is D_n (for some n) or E_8.

Clearly, only finitely many graphs are representable in E_8. So, to complete the proof, we only need to examine the case where the embedding is into D_n.

So suppose G is a connected graph with a Euclidean representation as a subset of D_n. We know that the vectors in the representation have non-negative inner products with each other. In particular, a vector and its negative cannot both occur.

Suppose that both $e_i + e_j$ and $e_i - e_j$ occur in the representation. Then no other vector can involve e_j, and none can involve $-e_i$. Thus we obtain a cocktail party graph consisting of vectors $e_i \pm e_k$ for some values of k. We call the vector e_i the *index* of this cocktail party.

If we have $-e_i + e_j$ and $-e_i - e_j$, then we can change the sign of e_i to obtain a cocktail party of the previous form with index e_i.

Any other basis vector e_l occurs in only one of $\pm e_l \pm e_m$, and we may assume that it occurs with positive sign.

Then the vectors consist of a number of cocktail parties, and some vectors $e_i + e_l$ where each of e_i and e_l is either the index of a cocktail party or one of the additional vectors (which could be regarded as indices of empty cocktail parties). So we have reconstructed the graph, and it is a generalized line graph. (In detail, take the vertices of the graph G to be the indices of the cocktail parties, including trivial indices of empty cocktail parties; edges of G have the form il, where $v_i + v_l$ is a vector in our set. Now we have precisely the representing set of the generalized line graph $L(G; a_1, \ldots)$, where $\mathrm{CP}(a_i)$ is the cocktail party whose index is the ith vertex of G.)

To illustrate the process, check that the generalized line graph shown in Fig. 9.1 is represented by the following vectors in D_{10}:

$$e_1 + e_2, e_1 + e_3, e_1 + e_4, e_2 + e_3, e_2 + e_4,$$

$$e_1 \pm e_5, e_1 \pm e_6, e_2 \pm e_7, e_4 \pm e_8, e_4 \pm e_9, e_4 \pm e_{10},$$

where the first 5 vectors represent the line graph and the remaining 12 the attached cocktail party graphs.

Many characterizations of graphs by their eigenvalues can be deduced from this theorem. Here is an example, a theorem of Hoffman and Ray-Chaudhuri. (Originally, the theorem was as follows: 'A sufficiently large connected regular graph ... is a line graph or a cocktail party graph.')

Theorem 9.6 *A connected regular graph whose adjacency matrix has least eigenvalue -2 or greater is a line graph, is a cocktail party graph, or is represented by a subset of E_8.*

Proof All we need to do is to show that a generalized line graph cannot be regular unless it is either a line graph or a cocktail party graph. So consider a generalized line graph $L(G; a_1, \ldots, a_n)$, which has an edge ij and has $a_i > 0$. Let d_i and d_j be the valencies of i and j in G. Then the vertex corresponding to the edge ij has valency $(d_i - 1) + (d_j - 1) + 2a_i + 2a_j$, whereas a vertex in $\mathrm{CP}(a_i)$ has valency $(2a_i - 2) + d_i$. These differ by $d_j + 2a_j$, where $d_j \geq 1$, a contradiction. So either G has one vertex and no edges (and we have just a cocktail party graph), or $a_i = 0$ for all i (and we have a line graph).

The regular graphs represented by subsets of E_8 have been classified by Bussemaker, Cvetković and Seidel [28]. There are 187 such graphs which are not line graphs or cocktail party graphs.

Another example is an even earlier theorem of Seidel [91] (which, however, included a classification of all the exceptional graphs which are strongly regular: there are exactly seven of them, the smallest being the Petersen graph). Seidel's theorem extended earlier results by Hoffman, Chang and Shrikhande.

Theorem 9.7 *A strongly regular graph whose adjacency matrix has least eigenvalue −2 is isomorphic to* $L(K_m)$, $L(K_{m,m})$ *or* $\mathrm{CP}(m)$, *or is one of finitely many exceptions.*

To prove this, we have to show that, if the line graph $L(G)$ is strongly regular, then G is complete, complete bipartite or a 5-cycle. (The 5-cycle is isomorphic to its own line graph; this can be put in with the finitely many exceptions.)

In fact, Seidel found the list of exceptions, with their parameters, which is as follows:

- the 5-cycle, $(5,2,0,1)$;
- the Petersen graph, $(10,3,0,1)$;
- the complement of $T(6)$, $(15,6,1,3)$;
- the complement of the Clebsch graph, $(16,10,6,6)$;
- the Shrikhande graph, $(16,6,2,2)$;
- the Schläfli graph, $(27,10,1,5)$;
- the three Chang graphs, $(28,12,6,4)$.

9.3 The Shrikhande Graph in E_7

As an example, consider the Shrikhande graph. Its adjacency matrix A has −2 with multiplicity 9, so $A + 2I$ has rank 7, and the graph is embedded in 7-dimensional space; since it is not a line graph or a cocktail party graph (Exercise 9.3), it can be realised as a subset of the root system E_7. We now give such a representation. We use the description of the Shrikhande graph given in Section 3.4.

We will use vectors orthogonal to $\frac{1}{2}(e_1 + \cdots + e_8)$ in the representation of E_8 given in Section 8.5. (Note that it doesn't matter which root of E_8 we choose here, since a suitable product of reflections will map any root to any other.) So all coordinates of vectors we use to build the Shrikhande graph must have equally many positive and negative entries. (Call such a root *admissible*.) We must choose 16 admissible roots such that all inner products are non-negative and the graph formed by joining two roots if they are not orthogonal is the Shrikhande graph.

We take the vertex $*$ to be $\frac{1}{2}(e_1 + e_3 + e_5 + e_7 - e_2 - e_4 - e_6 - e_8)$, which for brevity we write as $+1357-2468$. Now the vertices $1,\ldots,6$ of the hexagon are $e_1 - e_2$, $e_3 - e_2$, $e_3 - e_4$, $e_5 - e_4$, $e_5 - e_6$, $e_1 - e_6$. These are indeed admissible and joined to $*$, and joined only to their neighbours in the cycle. Label them $a,b,\ldots,f$.

Consider the root $e_1 - e_7$. It is not joined to $*$, and is joined only to $a = e_1 - e_2$ and $f = e_1 - e_6$ in the hexagon, so it corresponds to the edge af. Similarly $ab = e_8 - e_2$, $bc = e_3 - e_7$, $cd = e_8 - e_4$, $de = e_5 - e_7$, and $ef = e_8 - e_6$. We also see that two of these vertices are joined if and only if they both involve e_7 or both involve e_8; that is, they are two steps apart in the hexagon.

Finally, $ad = +1568-2347$, $be = +3458-1267$ and $cf = +1238-4567$. (For example, ad is joined to a, d, ab, af, cd and de.)

This completes the identification of a copy of the Shrikhande graph in the E_7 root system.

Exercises

9.1 Prove that the eigenvalues of the complete graph K_n are $n - 1$ (with multiplicity 1) and -1 (with multiplicity $n - 1$).

9.2 Show that a graph which has a Euclidean representation by a subset of the root system A_n is the line graph of a bipartite graph.

9.3 Show that the Shrikhande graph has least eigenvalue -2 but is not a line graph or a cocktail party graph.

9.4 Find a representation of the Petersen graph in the root system E_6.

9.5 Prove Theorem 9.7.

9.6 (a) Find a graph H such that the line graph of H is the complete graph K_n.

(b) Let G be obtained from the complete graph K_n by deleting m pairwise disjoint edges, where $2m < n$. Express G as a generalized line graph.

9.7 Let G be a graph with n vertices $v_1,\ldots,v_n$, and let $a_1,\ldots,a_n$ be non-negative integers. Show that $L(G; a_1,\ldots,a_n)$ is represented in the root system D_m, where $m = n + a_1 + \cdots + a_n$.

9.8 Prove that $L(K_4) = L(K_4; 0,0,0,0)$ is isomorphic to $CP(3) = L(K_1; 3)$.

Remark It is known that, if $n + a_1 + \ldots + a_n > 4$, then the generalized line graph $L(G; a_1,\ldots,a_n)$, where G is an n-vertex connected graph, determines G and $a_1,\ldots,a_n$ uniquely. This result is proved in [30]; the exceptions are related to the exceptional root system F_4. It generalizes the following theorem of Whitney [107]:

Theorem 9.8 *Let G and H be connected graphs. Then $L(G)$ is isomorphic to $L(H)$ if and only if either G is isomorphic to H, or $\{G,H\} = \{K_3, K_{1,3}\}$.*

9.9 Check that $L(K_3)$ and $L(K_{1,3})$ are isomorphic.

10

Miscellanea

In this chapter we examine other areas in which the Shrikhande graph has a role to play, including Seidel switching and equiangular line sets (which give rise to our final construction of the Shrikhande graph), design theory, Hadamard matrices and distance-regular graphs.

We begin with a short section indicating a few directions in which the study of the Shrikhande graph has been taken. Most of the detail is omitted, and we refer to the cited papers.

10.1 A Bouquet of Papers on the Shrikhande Graph

This section can be omitted at first reading. We simply want to show the variety of uses to which the Shrikhande graph can be put.

10.1.1 Deza Graphs

As we have seen, a regular graph is strongly regular if the number of common neighbours of two vertices v, w has just two values, depending on whether v and w are adjacent or not. Deza graphs generalize this.

A *Deza graph* with parameters (v, k, b, a), with $b \geq a$, is a regular graph of degree k on v vertices such that any two distinct vertices have either a or b common neighbours. It is called *strictly Deza* if its diameter is 2 (i.e., $a > 0$) and it is not strongly regular.

The definition comes from a major paper by Erickson *et al.* [46]. They gave the following construction:

Theorem 10.1 *Let G be a strongly regular graph with parameters (v, k, λ, μ), with $k \neq \mu \neq \lambda$, having adjacency matrix A. Let P be the permutation matrix*

125

corresponding to an automorphism α of G with the properties that α is an involution (that is, α^2 is the identity) and if (u, w) is a non-trivial cycle of α then u and w are non-adjacent.

Then PA is the adjacency matrix of a Deza graph with parameters v, k, $b = \max\{\lambda, \mu\}$ and $a = \min\{\lambda, \mu\}$.

This theorem does not apply to $L_2(4)$ and the Shrikhande graph, since these graphs have $\lambda = \mu$. However, Goryainov and Shalaginov [50] examined their complements and showed that, while the Shrikhande graph does not give rise to a Deza graph by this construction, the graph $L_2(4)$ does.

10.1.2 Quantum Automorphisms

The notion of quantum automorphism group was given by Banica [9] in 2005. It depends on the notion of the quantum symmetric group, a compact C^*-algebra given by generators and relations. It would take us rather far afield to define this.

Schmidt [90] showed that, of the smallest pair of strongly regular graphs with the same parameters, the quantum automorphism group coincides with the usual automorphism group for the Shrikhande graph, but $L_2(4)$ has extra quantum symmetries.

10.1.3 Locally Shrikhande Graphs

Makhnev and Paduchikh [74] examined graphs which are 'locally Shrikhande graphs', in the sense that the induced subgraph on any vertex neighbourhood is isomorphic to the Shrikhande graph. These graphs necessarily have either 40 or 80 vertices.

In a related work, Kardanova and Makhnev [62] considered graphs in which vertex neighbourhoods are the complements of strongly regular graphs with least eigehvalue -2, under extra hypotheses. One case they consider is graphs which are locally the complement of the Shrikhande graph.

Andries Brouwer has drawn our attention to a construction of the 80-vertex locally Shrikhande graph based on the work of Shepherd [93] in 1953.

Take the 4-dimensional vector space $\mathbb{C}^4$, where $\mathbb{C}$ is the field of complex numbers. In this space, let V_1 be the set of 16 vectors with ith coordinate $2r$ and other coordinates zero, where $r \in R = \{1, i, -1, -i\}$ and $1 \leq i \leq 4$; and V_2 the set of 64 vectors of the form (r_1, r_2, r_3, r_4) with $r_1, \ldots, r_4 \in R$ and $r_1 r_2 r_3 r_4 = 1$. Let $V = V_1 \cup V_2$ be this set of 80 vertices.

Note that R is the cyclic group of order 4 generated by i, so we can match it up with $\mathbb{Z}_4$, where $1 \leftrightarrow 0$, i $\leftrightarrow 1$, $-1 \leftrightarrow 2$, and $-$i $\leftrightarrow 3$.

Now identify $\mathbb{C}^4$ with $\mathbb{R}^8$, with orthonormal basis $\{e_j, ie_j : 1 \leq j \leq 4\}$. Let Γ be the graph with vertex set V, where two vertices are joined whenever their distance (in the Euclidean space) is 2. It can be checked that $2re_j$ and (r_1, r_2, r_3, r_4) are joined if and only if $r = r_j$, while (r_1, r_2, r_3, r_4) and (r'_1, r'_2, r'_3, r'_4) are joined if and only if they agree in two coordinates and the entries in each of the other two are not negatives of each other. For example, $(1,1,1,1)$ is joined to $(1, i, 1, -i)$. Now an easy check shows that the graph has valency 16.

We examine the induced subgraph on the neighbourhood of $2e_1$. We identify the vertex $(1, r_2, r_3, r_4)$ in this subgraph with the triple $(x_2, x_3, x_4) \in \mathbb{Z}_4^3$, where $r_j \leftrightarrow x_j$ as aforementioned. Now we have $x_2 + x_3 + x_4 = 0$, so the vertices are matched up with those of the Shrikhande graph as constructed in Section 2.6; we leave to readers the simple check that the edges also match up. Similarly the neighbourhood of any vertex in V_1 is the Shrikhande graph. We leave as an exercise a similar check for vertices in V_2.

10.1.4 Designs Formed by Independent Sets

Walikar *et al.* [104] study collections $\mathscr{B}$ of independent sets of maximum size in a graph G with the property that any two non-adjacent vertices lie in a constant number of members of $\mathscr{G}$. Among a number of examples they give is one in the Shrikhande graph, where the independent sets have size 4 and any two nonadjacent vertices lie in a unique block.

10.1.5 Generalized Shrikhande Graphs

These are graphs in which each pair of vertices have 0 or 2 common neighnours. Madani [73] constructed a family of these by taking hypercubes and adding proper diagonals.

10.2 Seidel Switching

In this section, we introduce Seidel switching, and show that this operation connects $L_2(4)$ to the Shrikhande graph. On the way, we see a few of the areas in which this concept has arisen.

10.2.1 Definition and Basic Properties

Jaap Seidel was an important figure in algebraic graph theory. One of his contributions was an operation on graphs called *Seidel switching* (to distinguish it from similar operations such as Godsil–McKay switching).

Let G be a graph on the vertex set V, and X a subset of V. The operation σ_X of *switching* G with respect to X involves the following:

- change each edge between a vertex in X and a vertex in $V \setminus X$ into a non-edge, and each such non-edge into an edge;
- leave the edges within or outside X unaltered.

The composition law for switching is easily verified:

$$\sigma_X \circ \sigma_Y = \sigma_{X \triangle Y},$$

where $X \triangle Y$ is the *symmetric difference* of X and Y. Note also that switching with respect to X and $V \setminus X$ are the same operation. So, if $|V| = n$, there are 2^{n-1} switching operations. Moreover, the composition law shows that switching gives an equivalence relation on the set of all graphs on the vertex set V, in which each equivalence class has size 2^{n-1}.

Seidel invented switching to study equidistant point sets in elliptic geometry, and applied it to the proof of Theorem 9.7 on strongly regular graphs with least eigenvalue -2. We will discuss this briefly before moving on to our central point, which is a simple construction of the Shrikhande graph by switching from the graph $L_2(4)$.

A brief digression. Seidel's thesis was, as suggested, on equidistant point sets in elliptic geometry. These are nothing but sets of lines through the origin in Euclidean space of one higher dimension, any two of which make the same angle. A set of points is *equidistant* if the distance between any pair of points is the same. It is easy to see that there cannot be more than three points in an equidistant set in the plane (realised by an equilateral triangle). But in the elliptic plane the maximum number is 6, and the configuration of 6 is given by the diagonals of a regular icosahedron (Fig. 10.1).

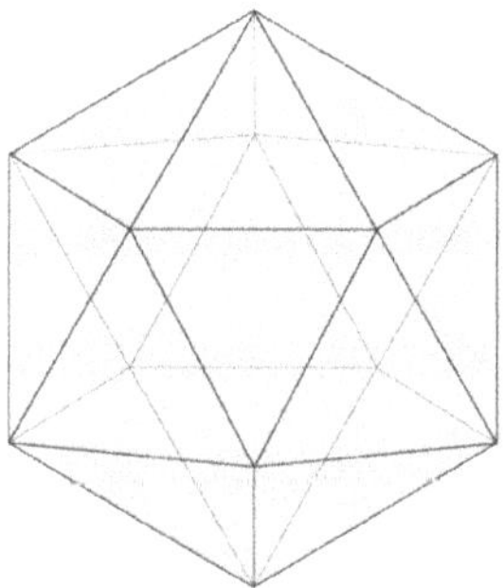

Figure 10.1 An icosahedron

To see that any two lines make the same angle, note that the vertices lie on a sphere, and so any two are at the same distance from the centre; thus two diagonals form two sides of an isosceles triangle whose base is the length of an edge of the polyhedron. The angle between two diagonals has cosine $1/\sqrt{5}$, as we will see shortly.

We can represent a set of equiangular lines in Euclidean space by a graph as follows. There is one vertex for each line. Choose a vector on each line; join two vertices if the corresponding vectors make an obtuse angle (more than a right angle).

If we make a different choice of directions on some of the lines, the effect on the graph is to apply Seidel switching. For example, changing the direction of a single line interchanges edges and nonedges between that vertex and all others. So the invariant attached to the set of lines is a switching class of graphs.

We define two more combinatorial objects.

- A *double cover of a complete graph K_n* is a graph on $2n$ vertices with the property that each vertex i of K_n corresponds to two vertices a_i and b_i of the double cover, and each edge of K_n to two edges forming a 1-factor between $\{a_i, b_i\}$ and $\{a_j, b_j\}$.
- A *two-graph* on a set V is a collection $\mathscr{T}$ of 3-element subsets of V with the property that any 4-element subset of V contains an even number of members of $\mathscr{T}$ (that is, 0, 2 or 4). Two-graphs were introduced by Graham Higman, who used them to give a combinatorial construction of the third sporadic group Co_3 found by John Conway.

Theorem 10.2 *The following objects are naturally equivalent:*

- *switching classes of graphs on V;*
- *sets of equiangular lines indexed by V;*
- *double covers of the complete graph on V;*
- *two-graphs on V.*

We will describe the correspondences between these four types of object, and leave the detailed verification to the reader.

We have already explained how a set of equiangular lines gives rise to a switching class of graphs. For the reverse direction, we associate a new matrix to a graph, called the *Seidel adjacency matrix*. If G is a graph on n vertices, the Seidel adjacency matrix B has rows and columns indexed by the vertices, with (i,j) entry 0 if $i-j$, -1 if vertices i and j are adjacent and $+1$ if they are distinct but non-adjacent. It is a simple exercise to show that, if the usual adjacency matrix of G is A, then $B = J - I - 2A$. From this it is possible to show that, if G is regular, then there is a simple relationship between the eigenvalues of A and B.

An important property of this matrix is as follows:

Proposition 10.3 *The spectrum of the Seidel adjacency matrix of a graph is unaffected by switching.*

Proof Suppose that the graph G has Seidel adjacency matrix B. Let X be a subset of the vertex set of the graph, and D the diagonal matrix with diagonal entries -1 in positions corresponding to points in X and $+1$ in the remaining diagonal positions. Then it is easily checked that the adjacency matrix of $\sigma_X(G)$ is $D^{-1}BD$, where B is the Seidel adjacency matrix of G. Now from elementary linear algebra, B and $D^{-1}BD$ have the same spectrum.

Suppose that the least eigenvalue of B is $-\alpha$. Then $B+\alpha I$ is positive semidefinite, and so is the Gram matrix of a set $v_1, \ldots, v_n$ of vectors in Euclidean space whose dimension is the rank of $B + \alpha I$. The diagonal entries are α, so each vector has length $\sqrt{\alpha}$; and the off-diagonal entries are ± 1, so distinct vectors have inner product -1 (if adjacent) and $+1$ (if non-adjacent). This means that the cosines of the angles between the lines spanned by these vectors are $\pm(1/\alpha)$; so the set of lines is equiangular.

In the other direction, if a set of lines is equiangular, making angles with cosines $\pm(1/\alpha)$, then choose vectors of length $\sqrt{\alpha}$ along each line; their Gram matrix is $\alpha I + B$, where B is a symmetric matrix with zero diagonal and off-diagonal entries ± 1, that is, the Seidel adjacency matrix of a graph. As noted earlier, a different choice of the directions of the vectors gives a switching-equivalent graph.

If we take vectors along the diagonals of the icosahedron in the hemisphere above the equator, then the central diagonal (the 'north pole') makes an acute angle with all the others; the remaining five make acute angles if and only if they are adjacent in the pentagon formed by the five edges which surround the north pole. So the Seidel adjacency matrix is

$$\begin{bmatrix} 0 & - & - & - & - & - \\ - & 0 & - & + & + & - \\ - & - & 0 & - & + & + \\ - & + & - & 0 & - & + \\ - & + & + & - & 0 & - \\ - & - & + & + & - & 0 \end{bmatrix}$$

Computer algebra finds that the characteristic polynomial of this matrix is $(x^2 - 5)^3$; so the eigenvalues are $\pm\sqrt{5}$, each with multiplicity 3. Thus, as expected, the switching class of this graph corresponds to a set of six equiangular lines in $\mathbb{R}^3$, with angle $\arccos(1/\sqrt{5})$ between any pair.

For the equivalence of switching classes and two-graphs, start with a graph G, and let $\mathscr{T}$ be the set of 3-element sets of vertices containing an even number of edges of G. Check that this is a two-graph, and is unaffected by switching. Conversely, given a two-graph on V and a point $v \in V$, define a graph with v as an isolated vertex, and an edge $\{x,y\}$ if and only if $\{v,x,y\} \notin \mathscr{T}$.

Finally, given a set of equiangular lines, choose unit vectors in *both* directions along all the lines as a vertex set, and join two of these vectors on different lines if their inner product is negative: this gives a double cover of the complete graph on V. In the other direction, given a double cover, choose one of each pair of vertices, and take the induced subgraph on the set of chosen vertices.

10.2.2 Regular Two-Graphs and Strongly Regular Graphs

A two-graph $\mathscr{T}$ on a set V is said to be *regular* if any two points v,w in V are contained in a constant number c of members of $\mathscr{T}$. A 4-subset of V is said to be *coherent* if all its 3-subsets belong to $\mathscr{T}$.

Lemma 10.4 *If $\mathscr{T}$ is a regular two-graph, then the number of coherent 4-sets containing a member of $\mathscr{T}$ is a constant d, with n = 3c − 2d.*

Proof Let $\{x,y,z\} \in \mathscr{T}$. Suppose that the number of coherent 4-sets containing $\mathscr{T}$ is d. For any point $w \notin \{x,y,z\}$, there are four possibilities: either $\{x,y,z,w\}$ is coherent, or the sets in $\mathscr{T}$ it contains are $\{x,y,z\}$ and one other; let the number containing $\{x,y,w\}$, $\{y,z,w\}$, $\{z,x,w\}$ be respectively p,q,r. Then counting arguments show that

$$d + p + q + r = n - 3, \quad d + p = d + q = d + r = c - 1.$$

So $2d = (3c − 3) − (n − 3) = 3c − n$, as required.

Now let Σ be a switching class of graphs on V. Then for $v \in V$, there is a unique graph $G_v \in \Sigma$ in which v is an isolated vertex: take any graph in Σ, and switch with respect to the neighbours of v.

Theorem 10.5 *Let Σ be a switching class of graphs on V, with corresponding two-graph $\mathscr{T}$. Then the following are equivalent:*

- *$\mathscr{T}$ is regular;*
- *$G_v \setminus \{v\}$ is strongly regular with $k = 2\mu$ and $n = 3k − 2\lambda$, for some $v \in V$;*
- *$G_v \setminus \{v\}$ is strongly regular with $k = 2\mu$ and $n = 3k − 2\lambda$, for all $v \in V$.*

Proof We do some counting. Let G be a graph on $n-1$ vertices; add an isolated vertex $*$, and count the triples on n vertices containing an odd number of edges of the resulting graph, through a given pair of vertices.

Case 1: $\{*,v\}$. Since $*$ is isolated, we count just the edges through v, and get the valency of v. So, if the two-graph is regular, then G is a regular graph, of valency, say, k.

Case 2: $\{v,w\}$ is an edge of G. Suppose that λ vertices of G are joined to v and w. Then the number joined to v but not w, or to w but not v, is $k - \lambda - 1$, and the number joined to neither is $n - 1 - 2 - 2(k - \lambda - 1) - \lambda$. So the number of odd triples containing $\{v,w\}$ is

$$1 + \lambda + n - 1 - 2k + \lambda = n - 2k + 2\lambda.$$

So regularity of $\mathscr{T}$ implies that λ is constant.

Case 3: $\{v,w\}$ is a non-edge. Suppose that μ vertices are joined to both. Then the number of odd triples is $2(k - \mu)$, so again regularity implies that μ is constant.

Moreover these three numbers must be equal:

$$k = n - 2k + 2\lambda = 2k - 2\mu,$$

so in particular $k = 2\mu$ and $n = 3k - 2\lambda$.

Conversely, if G is strongly regular with $k = 2\mu$, the arguments reverse to show that $\mathscr{T}$ is a regular two-graph, with associated parameter $c = k$.

We also need to know when a strongly regular graph lies in the switching class of a regular two-graph.

Theorem 10.6 *Let G be a strongly regular graph with parameters (n,k,λ,μ). Then the switching class of G corresponds to a regular two-graph if and only if $n = 2(2k - \lambda - \mu)$. The parameter c of the regular two-graph is given by*

$$c = n - 2k + 2\lambda = 2k - 2\mu.$$

Very similar counting arguments show that an edge lies in $n - 2k + 2\lambda$ 3-sets with an odd number of edges, while a non-edge lies in $2k - 2\mu$ such sets.

10.2.3 $L_2(4)$ and the Shrikhande Graph

Now after this rather long journey, we come to a beautiful vantage point.

Both $L_2(4)$ and the Shrikhande graph are strongly regular, with parameters $(16,6,2,2)$. Since $16 = 2(12 - 2 - 2)$, these graphs satisfy the condition of Theorem 10.6, and so both correspond to regular two-graphs with $c = 6$. From Theorem 10.5, we see that the graph obtained from either of them by isolating a vertex is strongly regular, with parameters $(15,8,4,4)$. The complement of this graph has parameters $(15,6,1,3)$. So it has the *strong triangle property*,

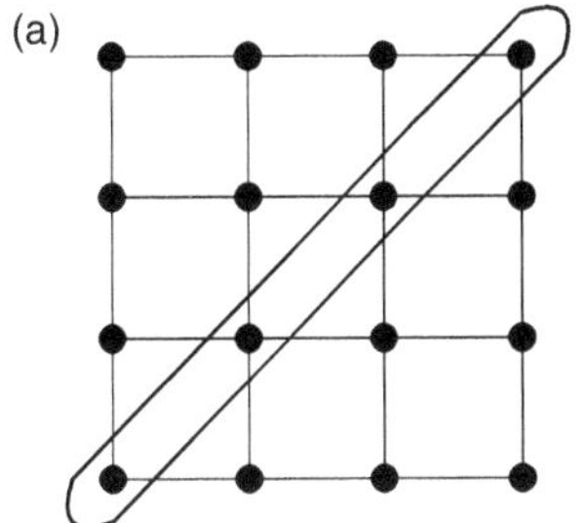 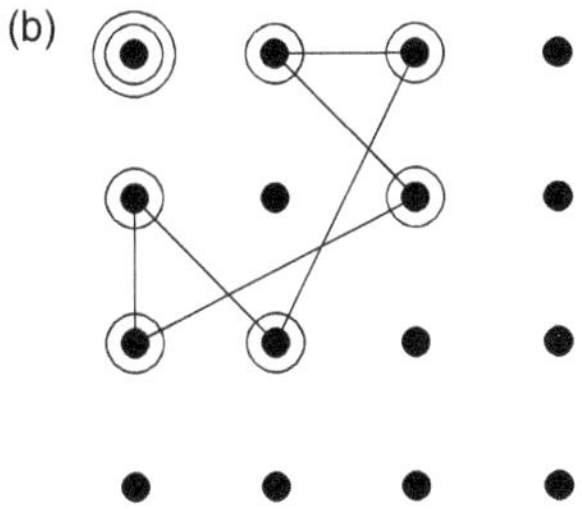

Figure 10.2 Switching set $L_2(4) \to$ Shrikhande (a) and a neighbourhood in the switched graph (b)

and we know from Theorem 3.4 or Theorem 9.7 that it is the complement of $T(6)$. Thus the two-graphs are the same, and we conclude:

Theorem 10.7 *The graph $L_2(4)$, the Shrikhande graph and the graph $T(6)$ with an isolated vertex all belong to the same switching class.*

Indeed, a small amount of trial and error shows that we can take the switching set to be a transversal to rows and columns in $L_2(4)$, as shown in Fig. 10.2. This gives us another construction of the Shrikhande graph: take the graph $L_2(4)$ and switch with respect to the transversal shown. In the second diagram, the singly circled vertices are the neighbours of the doubly circled vertex in the switched graph, with the edges between them shown; the neighbourhood of the doubly circled vertex thus induces a 6-cycle.

Finally, we remark that the complement of the Clebsch graph, with parameters $(16, 10, 6, 6)$ belongs to the same switching class, so it can be switched into $L_2(4)$ or the Shrikhande graph.

It can be shown that there are at most two possible parameter sets for a strongly regular graph in the switching class of a regular two-graph. In the case we have described, the two are both realized, one by $L_2(4)$ and the Shrikhande graph, the other by the complement of the Clebsch graph. In the switching class of the Schläfli graph with an isolated vertex, there are four different graphs with parameters $(28, 12, 6, 4)$, namely $T(8)$ and the three Chang graphs. The other possible parameter set is not realized (its complement would have parameters $(28, 9, 0, 4)$, and it can be shown that no such graph exists).

10.2.4 More on Switching Classes

We have seen that the switching class of the Shrikhande graph is a particularly interesting object, containing among others the graph $L_2(4)$, the complement of the Clebsch graph and $T(6)$ with an isolated vertex. There is also some research on switching classes in general, which we outline here. Our treatment is somewhat sketchy; we refer to the original papers for further details.

Enumeration

While collecting data for an earlier version of the On-Line Encyclopedia of Integer Sequences, Mallows and Sloane [76] noted that the number of isomorphism classes of switching classes on n points is equal to the number of even graphs (graphs in which each vertex has even degree, equivalently, graphs in which each connected component is Eulerian) on n vertices.

For odd n, there is a simple reason for this. If n is odd, then each switching class of graphs on n vertices contains a unique even graph: select any graph in the switching class, and switch it with respect to the set of vertices of odd degree. For even n, there is no such simple proof.

Mallows and Sloane proved the result by finding a formula for the number of switching classes, and noting that it is equal to a formula previously known for the number of even graphs. A feature of their proof is a beautiful observation: any automorphism of a switching class must fix some graph in the switching class.

The first author found a more conceptual proof. Both the set of all switching classes on an n-element set and the set of all even graphs on the same set have the structure of vector spaces over the 2-element field. In each case the operation is symmetric difference of the edge sets. (In the case of switching classes, we have to show that, if C_1 and C_2 are switching classes, then the set of all symmetric differences of a graph in C_1 with a graph in C_2 is a single switching class.) Now it is possible to show that these two vector spaces are duals to each other.

Now isomorphism types of even graphs or of switching classes are orbits of the symmetric group S_n acting on the appropriate vector space. It follows from a result known as *Block's Lemma* that, if a finite group acts on a finite vector space V and on its dual V', then the numbers of orbits are equal.

A Classification by Automorphisms

We have mentioned the result of Mallows and Sloane, that an automorphism of a switching class fixes a graph in that class. However, this result does not extend to groups of automorphisms. The switching class on 6 points corresponding to the diagonals of the regular icosahedron contains just four isomorphism types of graphs: the pentagon with an isolated vertex, the triangle with a pendant edge at each vertex and the complements of these; none is fixed by the full automorphism group of the class, the alternating group A_5.

So we can classify switching classes into two types: those in which the full automorphism group of the switching class fixes a graph in the class and those in which it does not. Let us call these 1 and 2.

There is a similar, more subtle characterization. Recall the double cover of the complete graph associated with a switching class. It has an automorphism ζ which swaps the two vertices covering each point in the switching class. So, if Γ is the automorphism group of the switching class, then the automorphism group Δ of the double cover has a normal subgroup of order 2 generated by ζ, with the quotient group $\Delta/\langle\zeta\rangle$ being isomorphic to Γ. It may be that Δ is isomorphic to $\langle\zeta\rangle\times\Gamma$ (so that Γ acts on the double cover), or it may be that this does not happen. Call these types A and B. So we have potentially four types of switching class: 1A, 1B, 2A, 2B.

Now it can be shown by methods of homological algebra that type 1B cannot occur, so we are left with a three-way classification.

It is known that type 2 occurs only if n is even. (We saw earlier that, if n is odd, a switching class contains a unique even graph, which is fixed by all automorphisms of the switching class). Moreover, type B only occurs if n is divisible by 8.

Heavily Covered Points

A result of Boros and Füredi (for $d = 2$) and Bárány (for arbitrary d) asserts that there is a constant c_d such that, given any set P of n points in d-dimensional Euclidean space $\mathbb{R}^d$, there is a point of the space lying in at least $c_d\binom{n}{d+1}$ simplexes with vertices in P (that is, a fraction at least c_d of all such simplexes). The problem, of course, is to find the optimal value of c_d. It is known that, for $d = 2$, the optimum is $\frac{2}{9}$, but no larger values are known precisely.

An improvement, using a method of Gromov, is given by Kral', Mach and Serenyi [67]. They use a result on switching classes, as follows.

A graph is called *Seidel-minimal* if it has the smallest number of edges of any graph in its switching class. This is equivalent to saying that, given any partition of the vertex set into two parts, there are at least as many non-edges as edges having vertices in different parts of the partition (if this failed, then switching with respect to a part would decrease the number of edges). Now there is a function f such that, if the Seidel-minimal graph in a switching class has density at least α, then the corresponding two-graph has density at least $f(\alpha)$. The authors use *flag algebras* to improve known bounds for f, and use these to improve the lower bounds for c_d for all d in the geometric problem.

10.3　Distance-Regular Graphs

Another class of graphs in whose theory the Shrikhande graph plays a role consists of the *distance-regular graphs*.

Definition　Let G be a connected graph of diameter d. Then G is *distance-regular* if there exist integers c_i, a_i, b_i such that, if the distance from x to y is i, then the number of vertices adjacent to y at distances $d - 1$, d and $d + 1$ respectively from x is c_i, a_i and b_i respectively. (Mnemonic: c = 'closer', b = 'beyond'.)

Definition　Let G be a connected graph of diameter d. Then G is *distance-transitive* if, whenever x, y, x', y' are vertices with $d(x,y) = d(x',y')$, where d is the graph distance, there is an automorphism of G carrying x to x' and y to y'.

For comprehensive surveys of these classes of graphs we refer to [24, 41].

Note that distance-regular graphs of diameter 2 are the same as strongly regular graphs.

A distance-transitive graph is distance-regular. For, if $d(x,y) = i = d(x',y')$, then an automorphism carrying (x,y) to (x',y') maps the neighbours of y which have distance $d - 1$, d and $d + 1$ to x respectively to the sets similarly defined for x', y'. The converse is false, and indeed the Shrikhande graph is the smallest example of a distance-regular graph which is not distance-transitive.

An important class of distance-regular graphs comprises the *Hamming graphs*. The Hamming graph $H(n,q)$ is the Cartesian product of n copies of the complete graph K_q; see Section 2.2. In other words, its vertices are all sequences of length n with entries from a fixed alphabet of size q, with two vertices adjacent if they differ in precisely one position.

Richard Hamming was a pioneer of information theory, and his graphs form a framework for the theory of error correction. We can think of the vertex set of the Hamming graph $H(n,q)$ as consisting of all words of length n whose letters are taken from an alphabet of size q. If a word is transmitted through a noisy communication channel, some of its entries may be changed into different letters. The number of such errors which have occurred (the *Hamming distance* between the transmitted and the received word) is equal to the graph distance between the two vertices of the Hamming graph.

So the strategy for error correction is to restrict the words transmitted to lie in a subset of the vertex set called a *code*. The *minimum distance* of a code is the smallest Hamming distance between two codewords. If the minimum distance of a code C is greater than $2e + 1$, and if at most e errors occur during transmission, then the received word resembles more closely the transmitted word than any other codeword. For, if c is the transmitted codeword, w the received word and c' any other codeword, then by the triangle inequality,

$$d(c,w) = d(c',w) \geq d(c,c') \geq 2e + 1,$$

and $d(c,w) \leq d$; so $d(c',w) \geq e + 1 > d(c,w)$. Such a code is said to *correct e errors*.

Now the Hamming graph $H(n,q)$ is distance-regular, with $c_i = i$ and $b_i = (n - i)(q - 1)$ for $0 \leq i \leq n$.

Michael Doob [44] constructed another graph with the same parameters as $H(n,q)$ for $q = 4$. Doob's graph consists of the Cartesian product of copies of K_4 and the Shrikhande graph G. If there are a copies of K_4 and b of G, the number of vertices is 4^{a+2b}, and the parameters are the same as those of $H(a + 2b, 4)$.

Yoshimi Egawa [45] proved the analogue of Shrikhande's theorem for these graphs (using Shrikhande's theorem in the proof):

Theorem 10.8 *Let G be a distance-regular graph of diameter n having the same parameters as $H(n,q)$.*

(a) If $q \neq 4$, then G is isomorphic to $H(n,q)$.
(b) If $q = 4$, then G is a Doob graph (possibly $H(n,q)$ or the Shrikhande graph).

Another characterization is due to Coolsaet [39], who considered a class of graphs satisfying the following three properties:

(a) two vertices at distance at most 2 have exactly two common neighbours;
(b) if v, w are vertices at distance 2, then there are four vertices adjacent to w and at distance 2 from p.

These properties both hold in $L_2(4)$ and in the Shrikhande graph; and indeed, they hold in the distance-regular Hamming graphs over a 4-letter alphabet and the Doob graphs discussed later, as well as a unique distance-regular graph on 24 vertices. The local structure of such a graph (the induced subgraph on a vertex neighbourhood) is uniquely determined.

Koolen [66] also gave a local characterization of Doob graphs in the same year.

In view of the coding connection, it is interesting to note that there have been several recent papers on codes in the Doob graphs. In particular, Priyadarsini and Ayyagari [82] describe a cryptosystem based on the Shrikhande graph. It depends on the relation between this graph and a *Hadamard matrix* of order 16, which we will discuss further later.

10.4 Block Designs

In the first section of this section, we saw that the Shrikhande graph, $L_2(4)$, and the complement of the Clebsch graph all lie in the same Seidel switching class. In this section we discuss another structure which Shrikhande, $L_2(4)$ and Clebsch share.

We begin with an example.

10.4.1 An Example

Recall from Section 6.4.3 the definition of a projective plane. The smallest projective plane is the *Fano plane*, over the 2-element field: it has 7 points and 7 lines; each line has 3 points, and any two points are contained in one common line. It is shown in Fig. 10.3a.

A statistician is designing an experiment to compare seven different varieties of wheat. She has arranged for seven farms in different parts of the country to take part in the experiment; each will set aside three plots of land in a field for the experiment. Clearly it would not be a good idea to plant one variety on each farm, since there would be no way to tell whether any observed difference in yields was due to different varieties of seed or different conditions on the farms.

The statistician decides to use the Fano plane to design the experiment. In Fig. 10.3b, the columns show the farms and the numbers in the boxes are the varieties of seed to be planted in the plots on the farm. (There is one more step before the actual allocation is made, known as *randomization*; but we will not discuss that here.)

It is known that this design is *optimal* in the sense that the estimates of treatment differences are unbiased and as accurate as possible (that is, have smallest variance).

10.4.2 Design Theory

We begin with a warning on terminology. Both combinatorial and statistical design theorists use the unadorned word 'design', but with very different

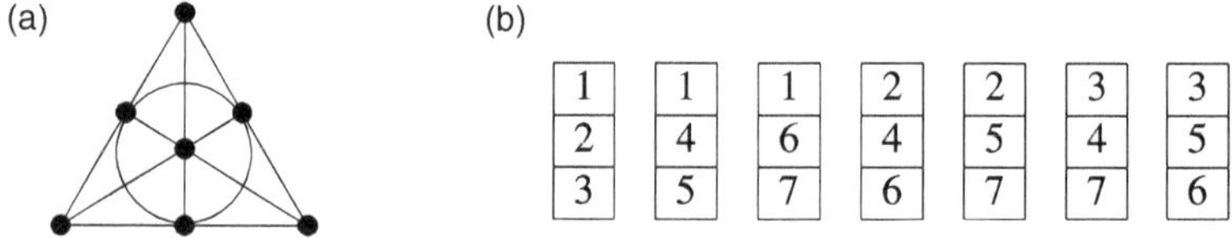

Figure 10.3 The Fano plane in geometry (a) and statistics (b)

meanings. To the statistician, the design is the function allocating treatments to experimental units. Usually the set of experimental units is structured in some way which will affect the experiment. This may be a partition into 'blocks' (such as the partition of plots into farms in our example), in which case we have a 'block design'. If the number of units in a block is smaller than the number of treatments, so that we cannot use every treatment in every block, we have an 'incomplete-block design'. If we can assign the treatments in such a way that any two treatments occur together in a block the same number of times, we have a 'balanced incomplete-block design', or BIBD for short; such a design, if it exists, is optimally efficient. For a combinatorial design theorist, the simple term 'design' usually means a BIBD; they often call such a design a *2-design*.

For detailed descriptions of the statistical and combinatorial viewpoints, we refer to the books [1] and [14]. We will take the combinatorialists' viewpoint, and begin with the definitions.

Definition Let v, k, λ be positive integers with $1 < k < v$. A $2\text{-}(v, k, \lambda)$ *design*, or *balanced incomplete-block design* (BIBD), consists of a set V of v points, together with a collection $\mathscr{B}$ of k-element subsets of V called blocks, such that any two points lie together in precisely λ blocks.

The use of the letter v here reflects the statistical origin of this concept: it is the initial letter of 'varieties'.

Thus the Fano plane is a $2\text{-}(7, 3, 1)$ design.

The next result gives some elementary properties.

Theorem 10.9 *Let $(X, \mathscr{B})$ be a $2\text{-}(v, k, \lambda)$ design.*

(a) Each point lies in $r = \lambda(v - 1)/(k - 1)$ blocks.
(b) Altogether there are $b = \lambda v(v - 1)/(k(k - 1))$ blocks.
(c) $b \geq v$.
(d) If $b = v$, then $r = k$ and any two blocks intersect in λ points.

Proof The first two parts are proved by double counting. For the first, choose a point x, and count pairs (y, B) with $y \neq x$ and $B \in \mathscr{B}$ in two different ways: choosing y first, there are $(v - 1)\lambda$ choices, while choosing B first, there are $r(k - 1)$ choices. For the second, a similar double count of pairs (x, B) with $x \in B \in \mathscr{B}$ shows that $vr = bk$; then, the first part is used.

The third part is the celebrated *Fisher's Inequality*. There is a counting proof, but the simplest proof uses linear algebra. Let M be the *incidence matrix* of the design, the $v \times b$ matrix with rows indexed by points and columns by blocks, with (x, B) entry 1 if $x \in B$, 0 otherwise. The definition of a 2-design shows that

$MM^\top = (r - \lambda)I + \lambda J$, where I and J are as usual the identity and all-1 matrices. Calculating eigenvalues of $(r - \lambda)I + \lambda J$ shows that this matrix is invertible, and so has rank v; thus M has at least as many columns as rows.

If equality holds, we have $r = k$ and M is square. Computing $MM^\top M$ and multiplying by the inverse of M shows that

$$M^\top M = (r - \lambda)I + \lambda J,$$

so any point lies in k blocks and any two blocks meet in λ points. This shows that the dual (with incidence matrix $M^\top$) is a design.

Conversely, if $D^\top$ is a design, then $b \geq v$ by Fisher's inequality, and $v \geq b$ by Fisher's inequality for $D^\top$; so $b = v$. $\qquad\square$

A design with $b = v$ is called a *symmetric 2-design* or *symmetric BIBD*.

Two constructions of new designs from old are *complement* and *dual*, as follows. Let $D = (V, \mathscr{B})$ be a 2-design.

(a) The *complement* is the design $D^c = (V, \{V \setminus B : B \in \mathscr{B}\})$; in other words, each block is replaced by its complement in V.
(b) The *dual* is the design $\mathscr{D}^\top = (\mathscr{B}, V^*)$, where V^* consists of all the sets $v^* = \{B \in \mathscr{B} : v \in B\}$; in other words, we interchange the role of points and blocks and reverse the incidence between them.

Theorem 10.10 (a) *If D is a 2-(v, k, λ) design, then D^c is a
2-$(v, v - k, b - 2r + \lambda)$ design.*
(b) *If D is a symmetric 2-design, then $D^\top$ is a symmetric 2-design with the same parameters.*

Proof For the first part, it is clear that the blocks of D^c have $v - k$ points; and given two points x and y; they are contained in λ blocks of D, $r - \lambda$ blocks contain x but not y and the same number of blocks contain y but not x; so there remain $b - 2r + \lambda$ blocks containing neither x nor y (and so the complementary blocks contain both).

The second part was proved earlier. $\qquad\square$

10.4.3 Designs from Graphs

Now we turn to a construction of designs from graphs. Recall that, in a graph G, the *neighbourhood* of a vertex v is $N(v) = \{w : w \sim v\}$, while the *closed neighbourhood* of v is $\bar{N}(v) = \{v\} \cup N(v)$.

Theorem 10.11 *Let G be a strongly regular graph with parameters n, k, λ, μ.*

(a) *If $\mu = \lambda$, then $(V(G), \{N(v) : v \in V(G)\})$ is a symmetric 2-(v, k, λ) design, with $v = n$.*

*(b) If $\mu = \lambda + 2$, then $V(G), \{\bar{N}(v) : v \in V(G)\}$ is a symmetric
2-$(v, k + 1, \lambda + 2)$ design, with $v = n$.*

The proof is straightforward. If $\mu = \lambda$, then any two vertices have λ common neighbours, and so lie in λ neighbourhoods. If $\mu = \lambda + 2$, then two non-adjacent points lie in μ closed neighbourhoods, while two adjacent points v, w lie in $\lambda + 2$ closed neighbourhoods, namely $\bar{N}(v)$, $\bar{N}(w)$, and each $\bar{N}(u)$ for $u \sim v, w$.

Corollary 10.12 *(a) Each of the graphs $L_2(4)$ and the Shrikhande graph has
the property that the neighbourhoods are the blocks of a symmetric
2-$(16, 6, 2)$ design.*
*(b) The closed neighbourhoods in the Clebsch graph are the blocks of a
symmetric 2-$(16, 6, 2)$ design.*
(c) These three designs are all isomorphic.

The first two parts follow immediately from the theorem, since $L_2(4)$ and the Shrikhande graph are strongly regular with parameters $(16, 6, 2, 2)$, while the Clebsch graph has parameters $(16, 5, 0, 2)$.

We will not prove the third part. It is known that there are precisely three symmetric 2-$(16, 6, 2)$ designs up to isomorphism; the one arising from our three strongly regular graphs has the largest automorphism group. It has another striking property.

First, we recall the definition of an affine geometry. If V is a vector space of dimension d over a field F, then the points of the geometry are the vectors in V, and the subspaces are the cosets of vector subspaces: so lines, planes, ..., hyperplanes are cosets of subspaces of dimension $1, 2, \ldots, d - 1$.

Theorem 10.13 *Let D be the symmetric 2-design from Corollary 10.12. Then*

*(a) the symmetric differences of pairs of blocks of D are the hyperplanes of
the 4-dimensional vector space over the binary field;*
*(b) the symmetric difference of three blocks is either a block or the
complement of a block.*

We will not give the proof of this theorem, but remark that it is probably easiest to take the graph $L_2(4)$. In this graph, one can show that

- if v_1, v_2, v_3, v_4 lie in different columns of the square array, then the symmetric difference of $N(v_1), \ldots, N(v_4)$ is empty, so that the symmetric difference of three of these sets is the fourth;
- if v_1, v_2 lie in the same row, then the symmetric difference of $N(v_1)$ and $N(v_2)$ is the union of the columns containing v_1 and v_2.

The second part of this theorem is known as the *symmetric difference property* for symmetric 2-designs.

10.4.4 Symplectic Designs

There is an infinite family of designs satisfying the symmetric difference property discussed in the preceding section. We now construct these; some background is required first.

Definition Let F be the two-element binary field, and V a vector space over F. A *bilinear form* on V is a function $\beta : V \times V \to F$ which is linear in each variable; that is,

- $\beta(ax + by, z) = a\beta(x,z) + b\beta(y,z)$,
- $\beta(x, ay + bz) = a\beta(x,y) + b\beta(x,z)$,

for all $x, y, z \in V$ and $a, b \in F$. Since $F = \{0, 1\}$, and these results are clear when the scalars are one, we can simplify this to two equations:

$$\beta(0,x) = \beta(x,0) = 0, \quad \beta(x + y, z) = \beta(x,z) + \beta(y,z),$$
$$\beta(x, y + z) = \beta(x,y) + \beta(x,z).$$

The form β is *non-degenerate* if

$$(\forall x \in V)(\beta(x,y) = 0) \Rightarrow x = 0,$$

and the corresponding equation in the second variable. Also, V is *alternating* if $\beta(x,x) = 0$ for all $x, y \in V$. Using bilinearity, expanding $\beta(x+y, x+y)$ and using $1 + 1 = 0$, we find that this implies that the form is symmetric: $\beta(x,y) = \beta(y,x)$ for all $x, y \in V$.

A *symplectic form* is a non-degenerate alternating bilinear form. From linear algebra, we learn:

Theorem 10.14 *If there is a symplectic form on V, then* $\dim(V)$ *is even, and the form is unique (up to change of basis).*

The standard alternating bilinear form is as follows: if $x = (x_1, \ldots, x_{2d})$ and $y = (y_1, \ldots, y_{2d})$, then

$$\beta(x,y) = (x_1 y_2 - x_2 y_1) + \cdots + (x_{2d-1} y_{2d} - x_{2d} y_{2d-1}).$$

Definition A *quadratic form* on V is a function $Q : V \to F$ satisfying

(a) $Q(ax) = a^2 Q(x)$;

(b) the function β given by

$$Q(x + y) = Q(x) + Q(y) + \beta(x,y)$$

is bilinear.

It is *non-degenerate* if β is.

In our case, with F the two-element field, the first condition simplifies to $Q(0) = 0$, and the second implies that

$$0 = Q(2x) = Q(x) + Q(x) + \beta(x,x) = \beta(x,x),$$

so β is alternating. We say that Q *polarizes* to β.

Theorem 10.15 *Let β be a symplectic form on a vector space V over the 2-element field F. Suppose that Q is a quadratic form which polarizes to β. Then the set of all quadratic forms polarising to β is*

$$\{Q + L : L \text{ is a linear form on } V\}.$$

Proof If Q and Q' both polarize to β, then $Q - Q'$ polarizes to 0, and thus is linear. Conversely, if we put $Q + L$ for Q in the polarization equation then L disappears. $\square$

Now we are ready to construct our designs.

Let F be the 2-element field, V a vector space of dimension $2d$ over F, β a symplectic form on V and Q a quadratic form which polarizes to β. Define $B_0 = \{x \in V : Q(x) = 0\}$, and for $y \in V$ let $B_y = B_0 + y$. Then $(V, \{B_y : y \in V\})$ is a symmetric 2-$(2^{2d}, 2^{2d-1} + \epsilon 2^{d-1}, 2^{2d-2} + \epsilon 2^{d-1})$ design.

Moreover, this design has the symmetric difference property: the symmetric difference of any three blocks is either a block or the complement of one. Moreover, the symmetric differences of pairs of blocks are the hyperplanes of the $2d$-dimensional affine space on V.

We will not prove this here.

10.5 Hadamard Matrices

The Shrikhande graph also gives a simple construction of a Hadamard matrix. We describe the context briefly and then exhibit the construction.

Hadamard asked the question: if a real square matrix has the property that all entries have modulus at most 1, how large can its determinant be? He showed:

Theorem 10.16 *Let $A = (a_{ij})$ be an $n \times n$ real matrix satisfying $|a_{ij}| \leq 1$ for all i,j. Then $|\det(A)| \leq n^{n/2}$.*

Proof The rows of A are n vectors in Euclidean space $\mathbb{R}^n$; each vector has Euclidean length at most $n^{1/2}$. The geometric interpretation of the determinant is that it is the n-dimensional volume of the parallelepiped whose edges are these n vectors. Clearly this volume is greatest when the vectors are pairwise orthogonal, and so the determinant is at most the product of the edge lengths, hence at most $n^{n/2}$. $\qquad\square$

When does equality hold?

Obviously, to have equality, each row must have Euclidean length $n^{1/2}$, so each entry a_{ij} must be $+1$ or -1; and the rows must be pairwise orthogonal, so that any two agree in $n/2$ positions and disagree in $n/2$ positions. This means that A is a matrix with entries ± 1 which satisfies

$$AA^\top = nI.$$

Such a matrix is called a *Hadamard matrix*.

The argument shows that, if $n > 1$, then n must be even for such a matrix to exist. One can go further:

Theorem 10.17 *If a Hadamard matrix of order n exists, then $n = 1$ or $n = 2$ or n is divisible by* 4.

Proof Suppose that $n > 2$, and consider three rows of A. Changing the signs of columns does not change the Hadamard property, so we may assume that the first row consists entirely of $+1$s. Let the first three rows be

$$
\begin{array}{cccc}
\overbrace{+\cdots+}^{a} & \overbrace{+\cdots+}^{b} & \overbrace{+\cdots+}^{c} & \overbrace{+\cdots+}^{d} \\
+\cdots+ & +\cdots+ & -\cdots- & -\cdots- \\
+\cdots+ & -\cdots- & +\cdots+ & -\cdots-
\end{array}
$$

Now orthogonality of the three pairs of rows gives

$$a + b = c + d,$$
$$a + c = b + d,$$
$$a + d = b + c$$

so $a = b = c = d$. Since $a + b + c + d = n$, we see that n is divisible by 4. $\qquad\square$

The *Hadamard conjecture* states that a Hadamard matrix exists for every order which is a multiple of 4. At the time of writing, the smallest multiple of 4 for which such a matrix is not known to exist is 668.

There are many known constructions. Our point here is simply to remark the following:

Proposition 10.18 *Let A be the adjacency matrix of either the graph $L_2(4)$ or the Shrikhande graph, and J the all-1 matrix. Then $2A - J$ is a symmetric Hadamard matrix.*

Proof We know that A is symmetric and $AJ = 6J$, $A^2 = 4I + 2J$. So $2A-$ has entries ± 1, and

$$(2A - J)^2 = 4A^2 - 4AJ + J^2 = 16I + 8J - 24J + 16J = 16I,$$

so the result is true. $\qquad\qquad\square$

For further reading on Hadamard matrices, we suggest the book by Horadam [60].

10.6 The Sylvester Graph

We conclude with a brief account of another beautiful graph, which has an indirect connection to the Shrikhande graph, and also with Shrikhande's other subject: Latin squares.

We have seen that, if n is a prime power, there exists a complete set of $n - 1$ mutually orthogonal Latin squares of order n. Thus, such complete sets exist for all integers n with $2 \le n \le 10$ except for $n = 6$. For this value, as conjectured by Euler, we cannot even find two orthogonal Latin squares. But what mathematics takes away with one hand, it gives with the other. There is a remarkable structure associated with $n = 6$, which allows the construction of a block design which is almost as good as the affine plane that would exist if there were a complete set of MOLS.

The structure involved is known as the *outer automorphism* of the symmetric group S_6, and 6 is the only number, finite or infinite, for which such a structure exists. The question is: given the symmetric group S_n on n points, does it have a non-trivial action on a set of size n which is essentially different from the action on the original n points, in the sense that the subgroups fixing points in the two actions are not conjugate?

In more combinatorial terms, we are asking: can we start with a set of n points and apply some combinatorial construction which builds another set of n objects, such that there is no natural matching between these n objects and the n original points? Sylvester discovered such a construction for $n = 6$, which we describe here. We will not give the proof that no such construction is possible for $n \ne 6$.

Sylvester was also a genius at inventing strange terminology; rather than following him, we will describe it in standard graph-theoretical terminology.

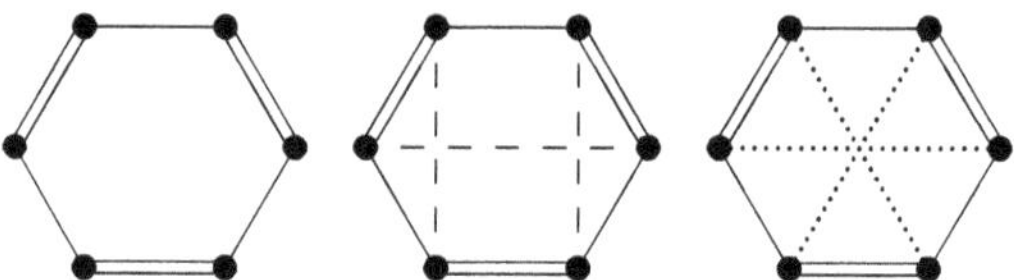

Figure 10.4 Two or three disjoint factors

We start with the complete graph K_6. A 1-factor is a set of three edges covering all the points, in other words, a spanning subgraph of valency 1. A 1-*factorization* is a partition of the 15 edges of K_6 into five 1-factors. For brevity, we will shorten these to *factor* and *factorization*.

Theorem 10.19 *There are exactly* 6 *distinct factorizations of* K_6. *Any two of them share a unique factor. The automorphism group of a factorization has order* 120 *and acts transitively on the vertices of* K_6.

Proof Consider the union of two factors. This is a graph of degree 2 on 6 vertices, hence a union of cycles; the factors alternate around any cycle, so all the cycle lengths are even, and there must be a single 6-cycle.

We have to add three more factors to construct a factorization. Inspection of a 6-cycle shows that its complement contains two distinct types of factor, a long diagonal and the two short diagonals perpendicular to it, and three long diagonals. Since there are 3 long diagonals and 6 short diagonals, we must use the three factors of the first type. This shows the uniqueness of the factorization. See Fig. 10.4.

Using this we can count the factorizations. There are 15 factors (each contains three edges, and each edge lies in three factors); and given one such factor, there are eight which are disjoint from it. Having chosen the first two, the factorization is unique. But in a given factorization, there are five choices of the first factor and four for the second. So the number of factorizations is

$$\frac{15 \cdot 8}{5 \cdot 4} = 6.$$

This argument also shows that two factorizations cannot have more than one factor in common. But there are 15 factors, and $\binom{6}{2} = 15$ pairs of factorizations; so any pair must have exactly one factor in common. $\qquad\square$

Using this construction, it is possible to build many combinatorial and geometric objects. Chapter 6 of the book [35] gives a number of these, including a projective plane, generalized quadrangle, Steiner system and the Hoffman–Singleton graph discussed in Section 3.2.1. In the Exercises, we give constructions of a projective plane of order 4 and of the Hoffman–Singleton graph.

Here we give a different construction: the Sylvester graph and a block design based on it. We note that Sylvester, a mathematician of wide interests, never considered this graph; it was named by Norman Biggs (personal communication) since it is based on Sylvester's construction in Theorem 10.19.

Let A be the set of vertices of K_6 and X the set of factorizations. We will take as the vertices of the Sylvester graph the set $A \times X$ of ordered pairs (a,x), where a is a vertex and x a factorization. Since all the factorizations are isomorphic and the automorphism group of one of them is transitive on the vertices, the symmetric group will act transitively on the 36 such pairs, and the graph we construct will be vertex-transitive.

We join two pairs (a,x) and (b,y) by an edge if $a \neq b$, $x \neq y$, and the edge $\{a,b\}$ belongs to the unique factor which x and y have in common.

Given a, x and y, the 1-factor shared by x and y has a unique edge $\{a,b\}$ containing a. Now, given a and x, there are five choices of $y \neq x$, and therefore five choices of a vertex (b,y) joined to (a,x). So the graph is regular with valency 5.

The vertex set $A \times X$ is best visualized as a 6×6 grid. Each of the five neighbours of a vertex lies in each of the five columns indexed by vertices $b \neq a$, and one in each of the five rows indexed by factorizations $y \neq x$. In fact it can be shown that the graph is *distance-transitive* as defined in Section 10.3. Two vertices at distance 2 lie in distinct rows and distinct columns (else the intermediate vertex would have two neighbours in the same row or column). So the 25 vertices in different rows and columns from (a,x) are the vertices at distance 1 or 2 from (a,x); and the remaining 10 vertices (in row x or in column a) are all at distance 3 from (a,x).

Now we proceed to the construction of a design, to replace the non-existent affine plane, following [4]. Call the 36 vertices of the graph *points*, and consider the following 48 subsets of size 6 of $A \times X$: the six rows and six columns of the array and the 36 closed vertex-neighbourhoods or starfish (Fig. 10.5). We call these subsets *blocks*.

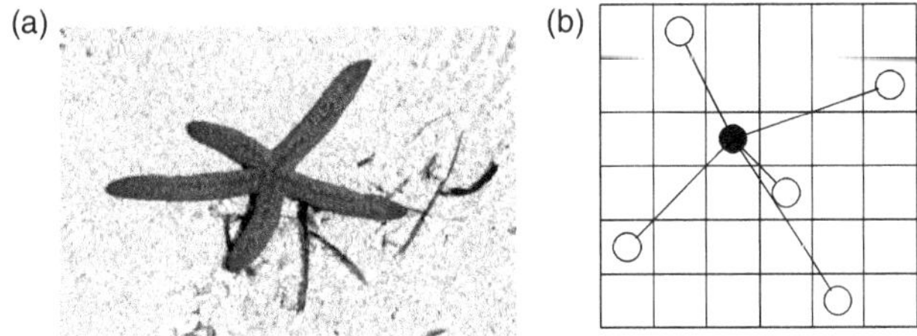

Figure 10.5 Starfish: on the Great Barrier Reef (a), and in the Sylvester graph (b)

Vertices *v* and *w* are contained in two blocks (the closed neighbourhoods of *v* and *w*) if *v* and *w* are joined in the Sylvester graph, and in exactly one block otherwise.

So, in a sense, this graph is not very far from being an affine plane (which would have every pair of points in a unique block). It shares another property of the affine plane: it is *resolvable*. This means that the blocks can be split into *parallel classes* each consisting of six blocks which partition the point set. This is a desirable property in experimental design. The parallel classes are the rows, the columns and the six *galaxies* of starfish, a galaxy consisting of the closed neighbourhoods of all the vertices in a column.

Also, it may be true that this design is the best possible for testing 36 treatments in 48 blocks of size 6 (in the sense that the random variables that give the best estimates of quantities we are interested in have the lowest possible variance, and hence the estimates are as accurate as possible), though this has not yet been proved.

10.7 A Final Note

We feel that while both Bose and Shrikhande worked in combinatorics and statistics (and so would have enjoyed the designs constructed here from the Sylvester graph), they were primarily mathematicians.

We note that, in the first few paragraphs of Section 10.4, we used the word 'field' with two entirely different meanings: the algebraic and the agricultural. In this connection (Fig. 10.6), we recall a story that was told at the memorial conference for Bose in 1988 [21].

Figure 10.6 Statistician and boss

There is a very famous joke about Bose's work in Giridh. Professor Mahalanobis wanted Bose to visit the paddy fields and advise him on sampling problems for the estimation of yield of paddy. Bose did not very much like the idea, and he used to spend most of the time at home working on combinatorial problems using Galois fields. The workers of the ISI used to make a joke about this. Whenever Professor Mahalanobis asked about Bose, his secretary would say that Bose is working in fields, which kept the Professor happy.

(The participants in Fig. 10.6 are not exact representations of Mahalanobis and Bose. We are grateful to Neill Cameron for the picture.)

Exercises

10.1 Complete the proof that the induced subgraphs on the neighbourhood of any vertex of the 80-vertex graph defined Section 10.1 are isomorphic to the Shrikhande graph.

10.2 Let Σ be the switching class corresponding to a pentagon with an isolated vertex. Show the following:

- Up to isomorphism, there are four different graphs in Σ. These consist of the pentagon with an isolated vertex, its complement (the wheel on six vertices), a triangle with one pendant edge at each vertex and the complement of this.
- The corresponding 2-graph is regular with parameter $a = 2$.
- The corresponding set of equiangular lines is the set of diagonals of a regular icosahedron.
- The corresponding double cover of K_6 is the 1-skeleton of the icosahedron.

10.3 In the graph $T(6)$, the line graph of K_6, a set of vertices corresponds to the edge set of a graph H on six vertices. Show that if we take H to consist of two disjoint triangles, then switching $\bar{T}(6) \cup \{*\}$ with respect to the edge set of H gives $L_2(4)$, while if H is a hexagon, we obtain the Shrikhande graph.

10.4 Show that there is no strongly regular graph with parameters $(28, 9, 0, 4)$.

10.5 (a) Show that switching the graph $L_2(4)$ with respect to the switching set shown in Section 10.2.3 really does produce the Shrikhande graph.

(b) Find a switching set in the complement of $T(6)$ with an isolated vertex which produces the Shrikhande graph.

10.6 Show that the double cover of the complete graph associated with a non-trivial regular 2-graph is distance-regular with diameter 3, and has a unique vertex at distance 3 from any given vertex.

10.7 Prove Theorem 10.13.

10.8 Show that the property of being a Hadamard matrix is unchanged if some rows and columns are multiplied by -1, if rows and columns are permuted or if the matrix is transposed. (Two Hadamard matrices related in this way are said to be *equivalent*.)

10.9 Let H_2 be the Hadamard matrix of order 2,

$$H_2 = \begin{pmatrix} + & + \\ + & - \end{pmatrix}.$$

Show that, if H is a Hadamard matrix of order n, then the Kronecker product

$$H \otimes H_2 = \begin{pmatrix} H & H \\ H & -H \end{pmatrix}$$

is a Hadamard matrix of order $2n$.

Deduce that $H_2 \otimes \cdots \otimes H_2$ (d factors) is a Hadamard matrix of order 2^d. (Such a Hadamard matrix is said to be of *Sylvester type*.)

10.10 (a) Show that the Hadamard matrix constructed from the Shrikhande graph is equivalent to a matrix of Sylvester type. (b) Is the same true for the matrix constructed from $L_2(4)$?

10.11 In the Sylvester graph, recall that a *galaxy* consists of all the starfish or closed vertex neighbourhoods corresponding to vertices in a fixed column. If we number the starfish in a galaxy from 1 to 6, and put the corresponding number in every cell of each starfish, show that we obtain a Latin square.

10.12 With the notation of Sylvester's construction, let $\mathscr{P}$ be the set of vertices and 1-factors of the complete graph K_6, and $\mathscr{L}$ the set of edges and 1-factorizations of K_6. Define incidence between $\mathscr{P}$ and $\mathscr{L}$ as follows: a vertex is incident with an edge containing it; an edge is incident with a 1-factor containing it; a 1-factor is incident with a 1-factorization containing it. Show that the incidence structure so constructed is a projective plane of order 4.

10.13 In this exercise we construct the Hoffman–Singleton graph.

Recall that the vertex set of the Sylvester graph is $A \times X$, where A is the set of vertices of K_6 and X the set of 1-factorizations. We take the vertex set of a graph G to be $\{\alpha, \xi\} \cup A \cup X \cup A \times X$, with edges as follows:

- $\{\alpha, \xi\}$;
- $\{\alpha, a\}$ for all $a \in A$;
- $\{\xi, x\}$ for all $x \in X$;
- $\{a, (a, x)\}$ and $\{x, (a, x)\}$ for all $a \in A$, $x \in X$;

- the edges of the Sylvester graph on $A \times X$.

Show that the constructed graph is strongly regular with parameters $(50, 7, 0, 1)$; that is, it has 50 vertices, degree 7, diameter 2 and girth 5.

10.14 Consider a galaxy of starfish (those whose centres lie in a fixed column of the 6×6 grid, without loss of generality the first). Take six letters, and put each in the positions of one starfish in the galaxy. Show that we obtain a Latin square whose automorphism group is the symmetric group S_5.

11

Further Reading

Through the window of the Shrikhande graph, we have introduced you to many important topics in discrete mathematics. We hope that you have been inspired to learn more about some of the topics, so we provide a list of books that go further than we have done. For a textbook on discrete mathematics, somewhat similar in spirit to this book, we recommend van Lint and Wilson [71].

For general graph theory, we suggest the books by Bollobás [18] and Diestel [42].

A series of books edited by Lowell Beineke and Robin Wilson [10, 11, 12] give a good overview of several aspects of graph theory: algebraic, topological, structural and chromatic [13].

Algebraic graph theory covers both linear algebra (spectral theory of graphs) and group theory (automorphism groups of graphs). We recommend the book by Brouwer and Haemers [25] for spectral graph theory. For permutation groups, look at [31] or [43]. The theory of maps on surfaces is dealt with in the book by Mohar and Thomassen [79]; see also Gross and Tucker [53]. The book by Krebs and Shaheen [68] covers Cayley graphs, and also the important graph-theoretic topic of expander graphs. The standard reference on distance-regular graphs is the book by Brouwer, Cohen and Neumaier [24]. Wilson [109] describes the famous Four-Colour Conjecture and its proof.

Ramsey's theorem, and the theory which has grown out of it, is described in the book by Graham, Rothschild and Spencer [51].

The standard reference on Latin squares is [63], a little dated now. There has been much recent activity in this area, and we await a new book describing what has been done.

For coding theory, a good introductory book, covering information and cryptography as well as coding theory, is the book by Biggs [15].

For design theory, as we explained in the text, there is a big difference between the views of combinatorialists and statisticians. For the former, the book [14] by Beth, Jungnickel and Lenz is a complete account; for the latter, see Bailey [1].

Finally, the ADE diagrams have a remarkable range of occurrences throughout mathematics, well beyond their appearance in spectral graph theory: they occur from finite groups to general relativity, from Lie algebras to singularity theory. The recent book [33] discusses these, and some of the links between them, though some mystery remains in explaining their ubiquity.

References

[1] R. A. Bailey, *Design of Comparative Experiments*, Cambridge University Press, Cambridge, 2008.

[2] R. A. Bailey and Peter J. Cameron, Laplacian eigenvalues and optimality, in *Groups and Graphs, Designs and Dynamics* (ed. R. A. Bailey, Peter J. Cameron and Yaokun Wu), pp. 176–265, London Math. Soc. Lecture Notes **491**, Cambridge University Press, Cambridge, 2024.

[3] R. A. Bailey, Peter J. Cameron, Cheryl E. Praeger and Csaba Schneider, The geometry of diagonal groups, *Trans. Amer. Math. Soc.* **375** (2022), 5259–5311.

[4] R. A. Bailey, P. J. Cameron, L. H. Soicher and E. R. Williams, Substitutes for the non-existent square lattice designs for 36 varieties, *J. Agric., Biol. Environ. Stat.* **25** (2020), 487–499.

[5] R. Balakrishnan, K. Ranganathan, *A Textbook of Graph Theory*, Springer, 2012. New York, NY, 2000.

[6] K. Balińska, D. Cvetković, Z. Radosavljević, S. Simić and D. Stevanović, A survey on integral graphs, *Publikacije Elektrotehničkog fakulteta. Serija Matematika* **13** (2002), 42–65.

[7] W. W. Rouse Ball, *Mathematical Recreations and Problems of Past and Present Times*, Macmillan, London 1982.

[8] H. J. Bandelt, H. M. Mulder, Distance-hereditary graphs, *J. Comb. Theory Ser. B* **41** (1986), 182–208.

[9] T. Banica, Quantum automorphism groups of homogeneous graphs, *J. Funct. Anal.* **224** (2005), 243–280.

[10] L. W. Beineke and R. J. Wilson (eds.), *Topics in Algebraic Graph Theory*, Cambridge University Press, Cambridge, 2004.

[11] L. W. Beineke and R. J. Wilson (eds.), *Topics in Topological Graph Theory*, Cambridge University Press, Cambridge, 2009.

[12] L. W. Beineke and R. J. Wilson (eds.), *Topics in Structural Graph Theory*, Cambridge University Press, Cambridge, 2013.

[13] L. W. Beineke and R. J. Wilson (eds.), *Topics in Chromatic Graph Theory*, Cambridge University Press, Cambridge, 2015.

[14] T. Beth, D. Jungnickel and H. Lenz, *Design Theory* (2nd edition), Cambridge University Press, Cambridge, 1999.

[15] N. Biggs, *Codes: An Introduction to Information Communication and Cryptography*, Springer-Verlag, London, 2008.

[16] N. L. Biggs, E. K. Lloyd and R. J. Wilson, *Graph Theory 1736–1936*, Oxford University Press, Oxford, 1976.

[17] G. D. Birkhoff and D. C. Lewis, Chromatic polynomials, *Trans. Amer. Math. Soc.* **60** (1946), 355–451.

[18] B. Bollobás, *Modern Graph Theory*, Springer, New York, 2025.

[19] Béla Bollobás and Andrew Thomason, Graphs which contain all small graphs, *Europ. J. Comb.* **2** (1981), 13–15.

[20] R. C. Bose, Strongly regular graphs, partial geometries and partially balanced designs, *Pacific J. Math.* **13** (1963), 389–419.

[21] Bose memorial session, in *Sankhyā* **54** (1992) (special issue devoted to the memory of Raj Chandra Bose), i–viii.

[22] R. C. Bose and S. S. Shrikhande, On the construction of sets of mutually orthogonal Latin squares and the falsity of a conjecture of Euler, *Trans. Amer. Math. Soc.* **95** (1960), 191–209.

[23] R. C. Bose, S. S. Shrikhande and E. T. Parker, Further results on the construction of mutually orthogonal Latin squares and the falsity of Euler's conjecture, *Canad. J. Math.* **12** (1960), 189–203.

[24] A. E. Brouwer, A. M. Cohen and A. Neumaier, *Distance-Regular Graphs*, Springer-Verlag, Berlin, 1989.

[25] A. E. Brouwer and W. H. Haemers, *Spectra of Graphs*, Springer, New York, 2012.

[26] R. H. Bruckinite nets. II: Uniqueness and imbedding, *Pacific J. Math.* **13** (1963), 421–457.

[27] R. H. Bruck and H. J. Ryser, The nonexistence of certain finite projective planes, *Canad. J. Math.* **1** (1949), 88–93.

[28] F. C. Bussemaker, D. M. Cvetković and J. J. Seidel, Graphs related to exceptional root systems, *Colloq. Math. Soc. Janos Bolyai* **18** (1978), 185–191.

[29] Peter J. Cameron, Cohomological aspects of two-graphs, *Math. Z.* **157** (1977), 101–119.

[30] P. J. Cameron, A note on generalized line graphs, *J. Graph Theory* **4** (1980), 243–245.

[31] Peter J. Cameron, *Permutation Groups*, London Math. Soc. Student Texts **45**, Cambridge University Press, Cambridge, 1999.

[32] Peter J. Cameron and Sebastian M. Cioabă, A graph partition problem, *Amer. Math. Monthly* **122** (2015), 972–983.

[33] P. J. Cameron, P.-P. Dechant, Y. H. He and J. McKay, *ADE: Patterns in Mathematics*, London Math. Soc. Student Texts **109**, Cambridge University Press, Cambridge, 2025.

[34] P. J. Cameron, B. Jackson and J. Rudd, Orbit-counting polynomials for graphs and codes, *Discrete Math.* **308** (2008), 920–930.

[35] P. J. Cameron and J. H. van Lint, *Graphs, Codes, Designs and Their Links*, London Math. Soc. Student Texts **22**, Cambridge University Press, Cambridge, 1991.

[36] P. J. Cameron and K. Morgan, Algebraic properties of chromatic roots, *Electron. J. Comb.* **24(1)** (2017), paper #P1.21.

[37] S. Chowla, P. Erdős and E. G. Straus, On the maximal number of pairwise orthogonal Latin squares of a given order, *Canad. J. Math.* **12** (1960), 204–208.

[38] Maria Chudnovsky, Neil Robertson, Paul Seymour and Robin Thomas, The strong perfect graph theorem, *Ann. Math.* **164** (2006), 51–229.

[39] K. Coolsaet, Local structure of graphs with $\lambda = \mu = 2$, $a_2 = 4$, *Combinatorica* **15** (1995), 481–487.

[40] D. G. Corneil, H. Lerchs and L. Stewart Burlingham, Complement reducible graphs, *Discr. Appl. Math.* **3** (1981), 163–174.

[41] E. R. van Dam, J. H. Koolen and H. Tanaka, Distance-regular graphs, Dynamic Survey, *Electron. J. Comb.*, version of 15 April 2016.

[42] Reinhard Diestel, *Graph Theory*, Springer, Heidelberg, 2017.

[43] J. D. Dixon and B. Mortimer, *Permutation Groups*, Springer, New York, 1996.

[44] Michael Doob, On graph products and association schemes, *Utilitas Math.* **1** (1972), 291–302.

[45] Yoshimi Egawa, Characterization of $H(n,q)$ by the parameters, *J. Comb. Theory, Ser. A* **31** (1981), 108–125.

[46] M. Erickson, S. Fernando, W. H. Haemers, D. Hardy and J. Hemmeter, Deza graphs: a generalization of strongly regular graphs, *J. Comb. Designs* **7** (1999), 359–405.

[47] The GAP Group, GAP – Groups, Algorithms, and Programming, Version 4.14.0; 2024. (www.gap-system.org)

[48] C. D. Godsil, GRR's for non-solvable groups, *Algebraic Methods in Graph Theory*, Vol. I, *Colloq. Math. Soc. Janos Bolyai* **25** (1981), 221–239.

[49] Chris Godsil and Gordon Royle, *Algebraic Graph Theory*, Springer, New York, 2001.

[50] S. V. Goryainov and L. V. Shalaginov, On Deza graphs with triangular and lattice graph complements as parameters, *J. Appl. Ind. Math.* **7** (2013), 355–362.

[51] Ronald L. Graham, Bruce L. Rothschild and Joel H. Spencer, *Ramsey Theory* (2nd edition), Wiley, New York, 1990.

[52] R. E. Greenwood and A. M. Gleason, Combinatorial relations and chromatic graphs, *Canad. J. Math.* **7** (1955), 1–7.

[53] Jonathan L. Gross and Thomas W. Tucker, *Topological Graph Theory*, Dover, New York, 2001 (reprint of Wiley edition of 1987).

[54] I. Gutman, O. E. Polansky, *Mathematical Concepts in Organic Chemistry*, Springer-Verlag, Berlin, 1986.

[55] Willem Haemers, Hoffman's ratio bound, *Linear Algebra Appl.* **617** (2021), 215–219.

[56] W. R. Hamilton, On some extensions of quaternions, *Proc. Royal Irish Acad.* **6** (1958), 114–115.

[57] R. Hammack, W. Imrich and S. Klavžar, *Handbook of Product Graphs*, Second Edition, CRC Press, Boca Raton, FL, 2011.

[58] A. J. Hoffman and R. R. Singleton, On Moore graphs with diameter 2 and 3, *IBM J. Res. Develop.* **5**(1960), 497–504.

[59] Derek Holton and John Sheehan, *The Petersen Graph*, Austral. Math. Soc. Lecture Series **7**, Cambridge University Press, Cambridge, 1993.

[60] K. J. Horadam, *Hadamard Matrices and Their Applications*, Princeton University Press, Princeton, 2007.

[61] Gareth A. Jones, Paley and the Paley graphs, in Gareth A. Jones, Ilia Ponomarenko and Jozef Širáň (eds.), *Isomorphisms, Symmetry and Computations in Algebraic Graph Theory*, Springer Proc. Math. Stat. **305** (2020), 155–183.

[62] M. L. Kardanova and A. A. Makhnev, On graphs in which the neighborhood of each vertex is the complementary graph of a Seidel graph, *Dokl. Math.* **82** (2010), 762–764.

[63] A. D. Keedwell and J. Dénes, *Latin Squares and Their Applications* (2nd edition), Elsevier, Amsterdam, 2015.

[64] T. P. Kirkman, Query VI, Lady's and Gentleman's Diary (1850), 48.

[65] Mikhail H. Klin and Andrew J. Woldar, The strongly regular graph with parameters $(100, 22, 0, 6)$: hidden history and beyond, *Acta Univ. M. Belii Ser. Math.*, **25** (2017), 5–62.

[66] J. H. Koolen, A characterization of the Doob graphs, *J. Comb. Theory, Ser. B* **65** (1995), 125–138.

[67] Daniel Král', Luáš Mach and Jean-Sébastien Sereni, A new lower bound based on Gromov's method of selecting heavily covered points, *Discrete Comput. Geom.* **48** (2012), 487–498.

[68] Mike Krebs and Anthony Shaheen, *Expander Families and Cayley Graphs*, Oxford University Press, Oxford, 2011.

[69] K. Kuratowski, Sur le problème des courbes gauches en topologie, *Fund. Math.* **15** (1930), 271–283.

[70] C. W. H. Lam, L. Thiel and S. Swiercz, The non-existence of finite projective planes of order 10, *Canad. J. Math.* **41** (1989), 1117–1123.

[71] J. H. van Lint and R. M. Wilson, *A Course in Combinatorics* (2nd edition), Cambridge University Press, Cambridge, 2012.

[72] L. Lovász, Normal hypergraphs and the perfect graph conjecture, *Discrete Math.* **2** (1972), 253–267.

[73] R. M. Madani, A class of minimal $\{0, 2\}$-graphs: "Generalized Shrikhande graphs", *Ars Comb.* **29C** (1990), 21–26.

[74] A. A. Makhnev and D. V. Paduchikh, Locally Shrikhande graphs and their automorphisms, *Sib. Math. J.* **39** (1998), 936–946.

[75] Roger Mallion, A contemporary Eulerian walk over the bridges of Kaliningrad, *BSHM Bull.* **23** (2008), 24–36.

[76] C. L. Mallows and N. J. A. Sloane, Two-graphs, switching classes, and Euler graphs are equal in number, *SIAM J. Appl. Math.* **28** (1975), 876–880.

[77] William S. Massey, *A Basic Course in Algebraic Topology*, Springer-Verlag, New York, 1991.

[78] A. Mohammadian, V. Trevisan, Some spectral properties of co graphs, *Discrete Math.* **339** (2016), 1261–1264.

[79] B. Mohar and C. Thomassen, *Graphs on Surfaces*, Johns Hopkins University Press, Baltimore, MD, 2001.

[80] E. J. F. Primrose, Kirkman's schoolgirls in modern dress, *Math. Gazette* **60** (1976), 292–293.

[81] Erich Prisner, *Graph Dynamics*, Chapman and Hall/CRC Press, Boca Raton, FL, 1995.

[82] P. L. K. Priyadarsini and Ramakalyan Ayyagari, Ciphers based on special graphs, in Proceedings 2013 International Conference on Advances in Computing, Communications and Informatics (ICACCI). Mysore, India.

[83] B. Radhakrishnan Nair, A. Vijayakumar, Strongly edge triangle regular graphs and a conjecture of Kotzig, *Discrete Math.* **158** (1996), 201–209.

[84] F. P. Ramsey, On a problem of formal logic, *Proc. London Math. Soc.* **30** (1930), 264–286.

[85] Stanisław Radziszowski, Small Ramsey numbers, *Electronic J. Combinatorics*, Dynamic Survey DS1.

[86] Suhail Ahmed Rather, Adam Burchardt, Wojciech Bruzda, Grzegorz Rajchel-Mieldzioć, Arul Lakshminarayan and Karol Życzkowski, Thirty-six entangled officers of Euler: quantum solution to a classically impossible problem, *Physical Rev. Lett.*, **128** (2022), 080507.

[87] Sharad S. Sane, The Shrikhande graph, *Resonance*, October 2015, 903–918.

[88] Sharad S. Sane, S S Shrikhande: the Euler Spoiler, *Resonance*, February 2021, 167–176.

[89] Sharad S. Sane, Reminiscences of a heritage: Professor Sharadchandra Shankar Shrikhande, *Math. Consortium Bull.* **2** (2020), 26–29.

[90] S. Schmidt, On the quantum symmetry of distance-transitive graphs, *Adv. Math.* **369** (2020), #107150.

[91] J. J. Seidel, Strongly regular graphs with $(-1,1,0)$ adjacency matrix having eigenvalue 3, *Linear Algebra Appl.* **1** (1968), 281–298.

[92] H. Seifert and W. Threlfall, *A Textbook of Topology*, Pure and Applied Mathematics, vol. 89, Academic Press, New York, NY, 1980.

[93] G. C. Shephard, Unitary groups generated by reflections, *Canad. J. Math.* **5** (1953), 364–383.

[94] M. S. Shrikhande, S. S. Shrikhande: his life and some contributions to mathematics, *Math. Student* **89** (3–4) (2020), 1–12.

[95] S. S. Shrikhande, The uniqueness of the L_2 association scheme, *Ann. Math. Stat.* **30** (1959), 781–798.

[96] N. M. Singhi and K. S. Vijayan, S. S. Shrikhande and his work: an appreciation, *J. Statist. Plan. Inference* **95** (2008), 3–7.

[97] N. M. Singhi and K. S. Vijayan, Sharad S. Shrikhande (1917–2020), *Curr. Sci.*, **118** (11) (2020), 1–3.

[98] Leonard H. Soicher, GRAPE: a GAP 4 package, Version 4.9.2, https://gap-packages.github.io/grape

[99] D. P. Sumner, Dacey graphs, *J. Australian Math. Soc.* **18** (4) (1914) 492–502.

[100] J. J. Sylvester, Chemistry and algebra, *Nature* **17** (1877–8), 284.

[101] W. T. Tutte, A contribution to the theory of chromatic polynomials, *Canad. J. Math.* **6** (1954), 80–91.

[102] Coen del Valle and Peter J. Dukes, Balancing permuted copies of multigraphs and integer matrices, *J. Comb. Theory, Ser. A* **198** (2023), 105756, 31 pp.

[103] V. G. Vizing, Some unsolved problems in graph theory, *Uspechi Mat. Nauk* **23** (1968), 117–134.

[104] H. B. Walikar, B. D. Acharya and S. S. Shirkol, Designs associated with maximum independent sets of a graph, *Designs Codes Cryptogr.* **57** (2007), 91–105.

[105] J. J. Watkins, *Across the Board: The Mathematics of Chessboard Problems*, Princeton University Press, Princeton, 2004.

[106] H. Whitney, A logical expansion in mathematics, *Bull. Amer. Math. Soc.* **38** (1932), 572–579.

[107] H. Whitney, Congruent graphs and the connectivity of graphs, *Amer. J. Math.* **54** (1932), 150–168.

[108] Richard M. Wilson, Concerning the number of mutually orthogonal Latin squares, *Discrete Math.* **10** (1974), 181–198.

[109] Robin Wilson, *Four Colours Suffice: How the Map Problem Was Solved*, Princeton University Press, Princeton, 2002.

Index